Lecture Notes in Bioinformatics 5109

Edited by S. Istrail, P. Pevzner, and M. Waterman

Subseries of Lecture Notes in Computer Science

Amos Bairoch   Sarah Cohen-Boulakia
Christine Froidevaux (Eds.)

Data Integration
in the Life Sciences

5th International Workshop, DILS 2008
Evry, France, June 25-27, 2008
Proceedings

 Springer

Series Editors

Sorin Istrail, Brown University, Providence, RI, USA
Pavel Pevzner, University of California, San Diego, CA, USA
Michael Waterman, University of Southern California, Los Angeles, CA, USA

Volume Editors

Amos Bairoch
University of Geneva, Swiss Institute of Bioinformatics
and Department of Structural Biology and Bioinformatics
CMU – 1, rue Michel Servet, 1211 Geneva 4, Switzerland
E-mail: amos.bairoch@isb-sib.ch

Sarah Cohen-Boulakia
Christine Froidevaux
Université Paris-Sud 11, Laboratoire de Recherche en Informatique
Bâtiment 490, 91405 Orsay Cedex, France
E-mail: {cohen, chris}@lri.fr

Library of Congress Control Number: Applied for

CR Subject Classification (1998): H.2, H.3, H.4, J.3

LNCS Sublibrary: SL 8 – Bioinformatics

ISSN   0302-9743
ISBN-10   3-540-69827-2 Springer Berlin Heidelberg New York
ISBN-13   978-3-540-69827-2 Springer Berlin Heidelberg New York

Springer is a part of Springer Science+Business Media

springer.com

© Springer-Verlag Berlin Heidelberg 2008
Printed in Germany

Typesetting: Camera-ready by author, data conversion by Scientific Publishing Services, Chennai, India
Printed on acid-free paper   SPIN: 12282746   06/3180   5 4 3 2 1 0

Preface

For several years now, there has been an exponential growth of the amount of life science data (e.g., sequenced complete genomes, 3D structures, DNA chips, mass spectroscopy data), most of which are generated by high-throughput experiments. This exponential corpus of data is stored and made available through a large number of databases and resources over the Web, but unfortunately still with a high degree of semantic heterogeneity and varying levels of quality. These data must be combined together and processed by bioinformatics tools deployed on powerful and efficient platforms to permit the uncovering of patterns, similarities and in general to help in the process of discovery. Analyzing complex, voluminous, and heterogeneous data and guiding the analysis of data are thus of paramount importance and necessitate the involvement of data integration techniques.

DILS 2008 was the fifth in a workshop series that aims at fostering discussion, exchange, and innovation in research and development in the area of data integration for the life sciences. Each previous DILS workshop attracted around 100 researchers from all over the world and saw an increase of submitted papers over the preceding one. This year was not an exception and the number of submitted papers increased to 54. The Program Committee selected 18 of them. The selected papers cover a wide spectrum of theoretical and practical issues including data annotation, Semantic Web for the life sciences, and data mining on integrated biological data.

Among these 18 papers, we distinguished 8 that describe research on new models, methods, or algorithms, and 8 that deal with the description of systems or experience with systems in practice. The two remaining papers have been selected for publication in a special issue of the *BMC Bioinformatics Journal.*

In addition to the presented papers, DILS 2008 featured three keynote talks by Olivier Bodenreider, National Library of Medicine, NIH, USA; Peter Karp, SRI International, USA; and Norman Paton, University of Manchester, UK. DILS 2008 also included a tutorial on bio-ontologies and a session dedicated to updates of biomolecular resources of world-wide importance: the UniProt knowledgebase and the EBI proteomics services.

The workshop was held at the University of Evry, in what is known as the 'Genomic Valley' at the heart of the Ile-de-France region, in France. DILS 2008 was kindly sponsored by the University of Paris-Sud 11, Microsoft Research who also made available their conference management system, the ENFIN network of Excellence, and the following institutes: IMGT, CEA, SIB, and CNRS (LRI and GDR BIM). We are very grateful to the University of Evry for hosting DILS, the MAISEL school for providing rooms for students, and the Genopole-Evry for its help in the local organization.

As editors of this volume, we thank all the authors who submitted papers, the Program Committee members and the external reviewers for their excellent work. Special thanks go to the local organizers, webmasters, Publicity and Sponsorship Chairs: Patrick Amar, Marie-Dominique Devignes, Nicole Lefèvre-Villain, Frédéric Lemoine, Isabelle Mougenot, Bastien Rance, Malika Smail, and Fariza Tahi. Finally, we are grateful for the cooperation of Springer in putting this volume together.

June 2008

Amos Bairoch
Sarah Cohen-Boulakia
Christine Froidevaux

Organization

Executive Committee

Program Chairs

Amos Bairoch	Swiss Institute of Bioinformatics, Swiss-Prot group, University of Geneva, Switzerland
Sarah Cohen-Boulakia	Laboratoire de Recherche en Informatique, CNRS UMR 8623, University of Paris-Sud 11, France
Christine Froidevaux	Laboratoire de Recherche en Informatique, CNRS UMR 8623, University of Paris-Sud 11, France

Program Committee

Amos Bairoch	SIB, University of Geneva, Switzerland
Sarah Cohen-Boulakia	LRI, University of Paris-Sud 11, France
Susan Davidson	University of Pennsylvania, USA
Marie-Dominique Devignes	LORIA, Nancy, France
Barbara Eckman	IBM, USA
Juliana Freire	University of Utah, USA
Christine Froidevaux	LRI, University of Paris-Sud 11, France
Floris Geerts	University of Edinburgh, UK
Amarnath Gupta	University of California San Diego, USA
Henning Hermjakob	EBI, UK
Ela Hunt	ETH Zurich, Switzerland
Minoru Kanehisa	Kyoto University, Japan
Jacob Koehler	University of Tromsø, Norway
Anthony Kosky	Axiope Inc., USA
Ulf Leser	Humboldt-Universität zu Berlin, Germany
Janice Lee Mong Li	National University of Singapore, Singapore
Frédérique Lisacek	SIB, University of Geneva, Switzerland
Bertram Ludäscher	University of California Davis, USA
Victor Markowitz	Lawrence Berkeley Labs, USA
Luc Moreau	University of Southampton, UK
Peter Mork	MITRE, USA
Fouzia Moussouni	INSERM, Rennes, France
Jignesh M. Patel	University of Michigan, USA

Manuel Peitsch	SIB, University of Basel, Switzerland
Erhard Rahm	University of Leipzig, Germany
Louiqa Raschid	University of Maryland, USA
Malika Smail	LORIA, Nancy, France
Val Tannen	University of Pennsylvania, USA
Thodoros Topaloglou	University of Toronto, Canada

External Reviewers

Sören Auer	Alasdair Gray	Timothy McPhillips
Jérôme Azé	Annika Gross	Anne Morgat
Shawn Bowers	Philip Groth	Krishna Palaniappan
I-Min Chen	Michael Hartung	Marie-Anne Poursat
Adrien Coulet	Samira Jaeger	Silke Trißl
Mohamed Elati	Toralf Kirsten	Patrick Ziegler
Isabelle Phan	Jérémie Mary	

Sponsoring Institutions

University of Paris-Sud 11	http://www.u-psud.fr/en/index.html/
Microsoft Research	http://research.microsoft.com/
ENFIN Network of Excellence	http://enfin.org/
IMGT	http://imgt.cines.fr/
CEA	http://www.cea.fr/
CNRS GDR BIM	http://www.gdr-bim.u-psud.fr/
CNRS LRI	http://www.lri.fr/
Swiss Institute of Bioinformatics	http://www.isb-sib.ch/

Sponsorship

Marie-Dominique Devignes	LORIA, Nancy, France
Malika Smail	LORIA, Nancy, France

Publicity

Julie Chabalier	Faculty of medicine, University of Rennes 1, France
Fouzia Moussouni	INSERM, Rennes, France

Webmasters

Frédéric Lemoine	LRI, University of Paris-Sud 11, France
Bastien Rance	LRI, University of Paris-Sud 11, France

Local Organization (France)

Patrick Amar   LRI, University of Paris-Sud 11
Sarah Cohen-Boulakia   LRI, University of Paris-Sud 11
Christine Froidevaux   LRI, University of Paris-Sud 11
Frédéric Lemoine   LRI, University of Paris-Sud 11
Isabelle Mougenot   LIRMM, University of Montpellier 2
Bastien Rance   LRI, University of Paris-Sud 11
Fariza Tahi   IBISC, University of Evry

DILS 2008 website   http://dils2008.lri.fr/

Table of Contents

Keynote Presentations

Semantic Web for the Life Sciences

Designing and Evaluating Architectures to Integrate Biological Data

New Architectures and Experience on Using Systems

Systems Using Technologies from the Semantic Web for the Life Sciences

Mining Integrated Biological Data

New Features of Major Resources for Biomolecular Data

DILS 2008 Tutorial

Ontologies and Data Integration in Biomedicine: Success Stories and Challenging Issues

Olivier Bodenreider

Lister Hill National Center for Biomedical Communications, National Library of Medicine,
National Institutes of Health, Bethesda, Maryland, USA
`olivier@nlm.nih.gcv`

Abstract. In this presentation, we review some examples of successful biomedical data integration projects in which ontologies play an important role, including the integration of genomic data based on Gene Ontology annotations, the cancer Biomedical Informatics Grid (caBIG) project, and semantic mashups created by the Semantic Web for Health Care and Life Sciences community.

1   Introduction

The promise of translational medicine hinges upon bridging basic research and clinical practice [1]. One key element to the integration of the research and clinical communities is the integration of the information sources and data used in these communities. In practice, bridges need to be created both across domains (e.g., between genotypic and phenotypic information sources) and across knowledge bases within a domain (e.g., between genomic and pathway resources). Biomedical ontologies play an important role in data integration [2]. They support data integration in two different ways, corresponding to two different approaches to data integration: warehousing and mediation [3]. One the one hand, by providing a controlled vocabulary in a given domain, ontologies support the standardization required from *warehousing approaches* to data integration, in which the sources to be integrated are transformed into a common format and converted to a common vocabulary. On the other hand, *mediation-based approaches* use ontologies for defining a global schema (in reference to which queries are made) and mapping between the global schema and local schemas (the schemas of the sources to be integrated).

We review examples in which ontologies have been used successfully for integrating biomedical data, including the integration of genomic data based on Gene Ontology annotations, the cancer Biomedical Informatics Grid (caBIG) project, and semantic mashups created by the Semantic Web for Health Care and Life Sciences community. Barriers to integration are discussed next.

2   Gene Ontology

The Gene Ontology (GO) [4] is a controlled vocabulary for the functional annotation of gene products across species [5]. In less than a decade, GO has been adopted by

A. Bairoch, S. Cohen-Boulakia, and C. Froidevaux (Eds.): DILS 2008, LNBI 5109, pp. 1–4, 2008.

several dozen model organism communities (e.g., Mouse Genome Informatics [6]) and has become a *de facto* standard for functional annotation. In addition to standardizing annotations across species, GO asserts relations among terms, which also facilitates data integration. GO is an enabling resource for comparative genomics, because it allows researchers to compare and contrast the functions of genes and gene products across multiple organisms [7]. Annotations repositories can be integrated not only with other annotation repositories, but also with a variety of data, including gene expression profiles (microarray data).

3   Cancer Biomedical Informatics Grid (caBIG)

The cancer Biomedical Informatics Grid (caBIG) of the National Cancer Institute (NCI) establishes a common infrastructure used to share data and applications across institutions to support cancer research efforts [8] in a grid environment [9]. Ontological resources such as the NCI Thesaurus [10] and the Cancer Data Standards Repository (caDSR) [11], a metadata registry for common data elements, are key resources of the common infrastructure for cancer informatics [12]. The data services currently available include, for example, caArray [13], a microarray data repository and gridPIR [14], a proteomic information resource based on UniProt and other databases from the Protein Information Resource (PIR). The Cancer Translational Research Informatics Platform (caTRIP) [15] takes a mediator-based approach to integrating a number of caBIG data services. Common data elements (CDEs) from the caDSR are used to join and merge data from the various repositories. CaBIG completed a 4-year pilot phase in 2007, involving 1,000 individuals from almost 200 organizations. In the next phase, caBIG tools and infrastructure will be made deployed to NCI-designated cancer centers.

4   Semantic Web for Health Care and Life Sciences

For the past two years, the World Wide Web Consortium (W3C) Health Care and Life Sciences Interest Group (HCLSIG) [16] has investigated the use of Semantic Web technologies in biomedicine. Ontologies play a central role in the Semantic Web [17], especially in biomedicine for which a large number of ontologies have been developed. This group advocates the use of Semantic Web technologies for supporting translational research [18] and has demonstrated the feasibility of integrating disparate resources in the domain of neurosciences, including Entrez Gene, Gene Ontology Annotations, the Allen Brain Atlas, PubMed/MEDLINE, and MeSH [19]. Other such "mashups" (integrative applications) have been developed since (e.g., [20]). Similar approaches have been used to integrate genotype and phenotype information [21], pathway and disease information [22], and to create drug-target networks [23]. Biomedical ontologies are crucial to these integration projects.

5   Challenging Issues

Freely and publicly available – preferably in several popular formats, easily discoverable and widely distributed ontologies are enabling resources for data integration, especially when they are embraced by active communities, used as a *de facto* standard in major data repositories and can interoperate with other ontologies. Integration is further facilitated by the availability of tools developed for and interfaces to these ontologies. This scenario essentially characterizes the Gene Ontology and explains in part its success.

There are, however, many obstacles preventing ontologies from being used efficiently for data integration. Despite the existence of repositories such as the National Center for Biomedical Ontology's BioPortal [24] and the Unified Medical Language System (UMLS) [25], not all ontologies can be accessed easily. Furthermore, some ontologies in the UMLS are subject to intellectual property restrictions and the UMLS cannot be used without first signing a license agreement. While OBO and OWL are popular formalisms for representing ontologies, many ontologies are only available in proprietary formats.

There is no authoritative mechanism for creating unique identifiers for biomedical entities. As a result, the same entity is often present under different identifiers in multiple ontologies, impeding integration. *Post hoc* mappings across ontologies such as those created by the UMLS somewhat alleviate this problem, but do not provide a complete solution. Additionally, in the Semantic Web, there is a need for a standard way of representing identifiers (e.g., URIs), as well as for services bridging identifiers across namespaces.

Differences in the granularity of annotations across datasets are also an issue, partially compensated by the use of aggregation strategies, such as the GO Slims [26] and the use of semantic similarity metrics [27]. Finally, not all datasets are directly amenable to integration. For example, metadata elements describing gene expression data in microarray repositories and fields in genome-wide association studies (e.g., Framingham Heart Study) are often in free text, not annotated to any ontology. Such datasets need to be preprocessed and encoded to an ontology prior to being integrated with other datasets.

References

1. Marincola, F.: Translational Medicine: A two-way road. Journal of Translational Medicine 1,1 (2003)
2. Bodenreider, O.: Biomedical ontologies in action: role in knowledge management, data integration and decision support. IMIA Yearbook of Medical Informatics, 67–79 (2008)
3. Hernandez, T., Kambhampati, S.: Integration of biological sources: Current systems and challenges ahead. Sigmod Record 33, 51–60 (2004)
4. Gene Ontology, http://www.geneontology.org/
5. Ashburner, M., Ball, C.A., Blake, J.A., Botstein, D., Butler, H., Cherry, J.M., Davis, A.P., Dolinski, K., Dwight, S.S., Eppig, J.T., Harris, M.A., Hill, D.P., Issel-Tarver, L., Kasarskis, A., Lewis, S., Matese, J.C., Richardson, J.E., Ringwald, M., Rubin, G.M., Sherlock, G.: Gene ontology: tool for the unification of biology. The Gene Ontology Consortium. Nat. Genet. 25, 25–29 (2000)
6. Mouse Genome Informatics, http://www.informatics.jax.org/
7. Blake, J.A., Bult, C.J.: Beyond the data deluge: data integration and bio-ontologies. J. Biomed. Inform. 39, 314–320 (2006)
8. caBIG Strategic Planning Workspace: The Cancer Biomedical Informatics Grid (caBIG): infrastructure and applications for a worldwide research community. Medinfo. 12, 330–334 (2007)

9. Oster, S., Langella, S., Hastings, S., Ervin, D., Madduri, R., Phillips, J., Kurc, T., Siebenlist, F., Covitz, P., Shanbhag, K., Foster, I., Saltz, J.: caGrid 1.0: an enterprise Grid infrastructure for biomedical research. J. Am. Med. Inform. Assoc. 15, 138–149 (2008)
10. NCI Thesaurus, http://www.nci.nih.gov/cancerinfo/terminologyresources
11. Cancer Data Standards Repository (caDSR), http://ncicb.nci.nih.gov/NCICB/infrastructure/cacore_overview/cadsr
12. Komatsoulis, G.A., Warzel, D.B., Hartel, F.W., Shanbhag, K., Chilukuri, R., Fragoso, G., Coronado, S., Reeves, D.M., Hadfield, J.B., Ludet, C., Covitz, P.A.: caCORE version 3: Implementation of a model driven, service-oriented architecture for semantic interoperability. J. Biomed. Inform. 41, 106–123 (2008)
13. caArray, http://caarray.nci.nih.gov/
14. gridPIR, https://cabig.nci.nih.gov/tools/PIR
15. caTRIP, https://cabig.nci.nih.gov/tools/caTRIP
16. Health Care and Life Sciences Interest Group, http://www.w3.org/2001/sw/hcls/
17. Schroeder, M., Neumann, E.: Semantic web for life sciences. Web Semantics: Science, Services and Agents on the World Wide Web 4, 167–167 (2006)
18. Ruttenberg, A., Clark, T., Bug, W., Samwald, M., Bodenreider, O., Chen, H., Doherty, D., Forsberg, K., Gao, Y., Kashyap, V., Kinoshita, J., Luciano, J., Marshall, M.S., Ogbuji, C., Rees, J., Stephens, S., Wong, G.T., Wu, E., Zaccagnini, D., Hongsermeier, T., Neumann, E., Herman, I., Cheung, K.H.: Advancing translational research with the Semantic Web. BMC Bioinformatics 8, Suppl. 3, S2 (2007)
19. HCLS Banff 2007 demo, http://esw.w3.org/topic/HCLS/Banff2007Demo
20. Sahoo, S.S., Bodenreider, O., Rutter, J.L., Skinner, K.J., Sheth, A.P.: An ontology-driven semantic mash-up of gene and biological pathway information: Application to the domain of nicotine dependence. Journal of Biomedical Informatics (2008), doi:10.1016/j.jbi.2008.1002.1006
21. Butte, A.J., Kohane, I.S.: Creation and implications of a phenome-genome network. Nat. Biotechnol. 24, 55–62 (2006)
22. Chabalier, J., Mosser, J., Burgun, A.: Integrating biological pathways in disease ontologies. Medinfo. 12, 791–795 (2007)
23. Yildirim, M.A., Goh, K.I., Cusick, M.E., Barabasi, A.L., Vidal, M.: Drug-target network. Nat. Biotechnol. 25, 1119–1126 (2007)
24. BioPortal, http://www.bioontology.org/tools/portal/bioportal.html
25. Bodenreider, O.: The Unified Medical Language System (UMLS): integrating biomedical terminology. Nucleic Acids Res. 32, 267–270 (2004)
26. GO Slim, http://www.geneontology.org/GO.slims.shtml
27. Lord, P.W., Stevens, R.D., Brass, A., Goble, C.A.: Semantic similarity measures as tools for exploring the gene ontology. In: Pac. Symp.Biocomput., pp. 601–612 (2003)

BioWarehouse: Relational Integration of Eleven Bioinformatics Databases and Formats

Peter D. Karp, Thomas J. Lee, and Valerie Wagner

SRI International, Menlo Park, CA USA

Abstract. BioWarehouse is an open-source project for integrating bioinformatics databases within a relational database warehouse. It has two key features. A comprehensive database schema models many different bioinformatics datatypes. A set of loader tools permits loading of public bioinformatics databases, and of standard bioinformatics formats, into that database schema. Thus, multiple databases can be queried together within a single common schema. The supported databases are BioCyc, CMR, ENZYME, Eco2DBase, Genbank, Gene Ontology, KEGG, NCBI Taxonomy, and UniProt. The supported formats are BioPAX (protein interactions subset only) and MAGE-ML.

1 Introduction

The BioWarehouse project [1,2,3] is pursuing a physical integration approach to the database integration problem for bioinformatics. In addition, BioWarehouse provides a relational database system for SQL query access to individual bioinformatics databases.

BioWarehouse can be implemented on Oracle and MySQL relational database management systems (DBMSs). Users can download the BioWarehouse software and set up a local implementation of BioWarehouse that contains loaded data of interest.

Why did we choose the warehouse approach instead of the multidatabase (federated) approach? The warehouse approach has the following advantages.

- The multidatabase approach assumes that the databases to be integrated are available in a queryable, network-accessible DBMS, which is often not the case in bioinformatics (such as for databases that are available only via downloadable files and/or as clickable web sites).
- Most sites that do provide their databases in a queryable DBMS do not allow remote query access because of security and loading concerns.
- Users often want to control data stability; they want to control when the data change so that they can perform reproducible experiments. In the multidatabase approach, data change at the discretion of the maintainers of the source data. On the other hand, the multidatabase approach does ensure access to the latest version of the data for those users that need such access.
- The bandwidth of the internet limits the throughput of querying and result returning.

A. Bairoch, S. Cohen-Boulakia, and C. Froidevaux (Eds.): DILS 2008, LNBI 5109, pp. 5–7, 2008.

- Users need to capture and integrate locally produced data of different types, which requires a local database.

2   The BioWarehouse Schema

The first component of BioWarehouse is a set of relational database schema definitions that model many bioinformatics datatypes. The schema is stored in a format that can be automatically converted to an Oracle schema, and a MySQL schema. The datatypes covered by the BioWarehouse schema include:

- Replicons, genes, and proteins
- Pathways, reactions, and small molecules
- Sequences and sequence features
- Controlled vocabularies
- Gene expression data
- Protein expression data
- Flow cytometry data
- Organisms and taxonomic relationships
- Results of computations, such as sequence matches
- Citations
- Links to external databases

An important aspect of the BioWarehouse approach is that data of the same type from different source databases is loaded into the same BioWarehouse tables. For example, protein data, be it from UniProt, KEGG, or BioCyc, is loaded into the same protein table within BioWarehouse. This approach allows all data of a given type to be queried together within the same tables.

The schema also models the source databases themselves. Every BioWarehouse object (e.g., a protein) is registered within the source database from which it was loaded. Multiple versions of a given dataset (e.g., KEGG) can be loaded side by side within a BioWarehouse instance.

3   The BioWarehouse Loaders

BioWarehouse loaders parse a source dataset, and load the contents of the dataset into appropriate BioWarehouse tables. For example, the BioCyc loader parses BioCyc data files describing genes, proteins, pathways, reactions, and small molecules, and load those data into a BioWarehouse instance. To date all loaders have been written in either the C or the Java language. We generally prefer to work with XML-format input files when they are available, because they are typically easier to parse than are other file formats invented by bioinformatics researchers.

BioWarehouse loaders exist for the following bioinformatics databases: Bio-Cyc, CMR, ENZYME, Eco2DBase, Genbank, Gene Ontology, KEGG, NCBI Taxonomy, and UniProt. BioWarehouse loaders exist for the following bioinformatics formats: BioPAX (protein interactions subset only) and MAGE-ML.

Documentation is provided for each BioWarehouse loader that specifies the format it accepts, and the data transformations it applies. That is, the documentation describes for fields in the input file, what columns in BioWarehouse tables those fields are mapped to.

4 Discussion

BioWarehouse is in active use at SRI for several projects. The BioCyc project generates high-throughput pathway predictions by loading hundreds of genomes from the CMR database into BioWarehouse, and then processing those genomes through a pathway prediction pipeline. The Pathway Tools project uses BioWarehouse to extract subsets of data from UniProt for use by the pathway hole filler component of Pathway Tools. Our enzyme genomics project has used BioWarehouse to find which enzymes with known biochemical activities have no associated sequence information.

We would be grateful for contributions of BioWarehouse extensions by the user community, such as new BioWarehouse loaders.

References

1. BioWarehouse home page, `http://bioinformatics.ai.sri.com/biowarehouse/`
2. Lee, T.J., Pouliot, Y., Wagner, V., Gupta, P., Stringer-Calvert, D.W.J., Tenenbaum, J.D., Karp, P.D.: BioWarehouse: a bioinformatics database warehouse toolkit. BMC Bioinformatics 7, 170 (2006)
3. Pouliot, Y., Lee, T.J., Wagner, V., Karp, P.D.: Identifying candidate genes using the biowarehouse: A case study. In: Proceedings of the Eighteenth International Conference on Systems Engineering, pp. 332–340. IEEE Computer Society, Los Alamitos (2005)

Data Integration in the Life Sciences: Fun, Findings and Frustrations

Norman W. Paton

School of Computer Science, University of Manchester
Oxford Road, Manchester M13 9PL, UK
npaton@manchester.ac.uk

Abstract. This paper concerns the research topic of data integration in the life sciences. The paper presents no technical results, but rather provides a classification of research activities in terms of the contributions they seek to make to the life sciences, bioinformatics or computer science.

1 Introduction

Research involving data integration in the life sciences is diverse in nature, being conducted by researchers with different backgrounds and objectives. Research can be classified into the five areas represented by the overlapping circles in Figure 1, which in turn can be characterised (left-to-right) as follows:

Life Science for its own sake: The use of informatics to obtain biological insights. Typically, where the aim is to obtain insight into some biological system or experimental method, existing informatics techniques are deployed. Results are published in the life sciences literature (e.g. [2]).

Bioinformatics for Life Science: The use of novel bioinformatics to learn specific biological lessons. Such an activity requires the development of a novel result in bioinformatics to enable a specific biological system or technique to be better understood. Results are typically published in the life sciences or computational biology literature (e.g. [7]).

Bioinformatics for its own sake: The development of novel generic (organism independent) bioinformatics techniques. Typically, the new technique is not widely applicable outside the life sciences, and results are not necessarily accompanied by new insights into biological systems. Results are typically published in the biotechnology or bioinformatics literature (e.g. [6]).

Bioinformatics for Computing: The use of the life sciences as a source of challenging computing problems. Results are typically published in the bioinformatics or computing literature (e.g. [3]).

Computing for its own sake: Computing research motivated by or illustrated using biological problems. Results are typically published in the computing literature (e.g. [1]).

The diverse range of types of result (from discoveries in the life sciences to generic techniques in computer science) from research under the heading of "data integration in the life sciences" has a number of implications for researchers working in the area, as discussed in the next section.

A. Bairoch, S. Cohen-Boulakia, and C. Froidevaux (Eds.): DILS 2008, LNBI 5109, pp. 8–10, 2008.

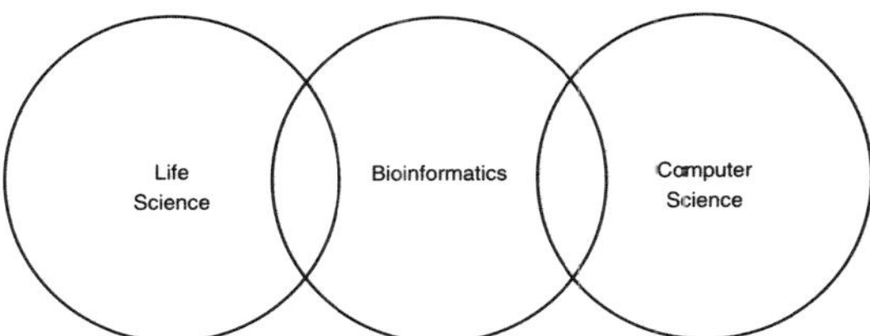

Fig. 1. Research areas of relevance to data integration in the life sciences

2 Observations

The following observations can be made about research on data integration in the life sciences:

Individuals: Few researchers are successful across the full range of areas in Figure 1, and those who are generally play a supporting role at one or both edges. This is neither surprising nor problematic, as it enables interdisciplinary teams to publish across the full spectrum.

Projects: Few projects are successful across the full range of areas in Figure 1, and those that are rarely apply the most novel computing when obtaining biological results. Indeed, individual projects typically occupy one or a few (adjacent) segments in Figure 1. This is not surprising, as deploying emerging computational techniques on applications that require dependable outcomes is a risky strategy. It may be considered problematic, however, as an interdisciplinary team that includes both computer and life scientists may be unlikely to generate research of direct interest to all its participants. Furthermore, the effectiveness of new computing techniques may not be subject to much practical evaluation in relevant applications.

The "Data Integration in the Life Sciences" (DILS) Workshop series is probably most naturally located in the *Bioinformatics for Computing* segment described in Section 1. If so, then the community is principally seeking to refine computational techniques for data integration in the light of challenges identified in life science applications. In common with other research in computer science, techniques under current investigation are not the finished article, and implementations are typically early prototypes or proofs of concept; as a consequence, results generated by this community are often not ready for large-scale deployment.

Overall, reflecting the focus on novel data integration techniques, there is little evidence of technical consolidation. The diversity of research reflects both differences in requirements in different areas of the life sciences, and the fact that various aspects of data integration in the life sciences are difficult in ways that are hard to address systematically. For example, independently developed and autonomously maintained data resources often provide integrators with rapidly changing models and interfaces, inconsistent descriptions of common concepts, incompatible identification schemes, etc. Such features make high-quality data integration solutions (e.g. through warehouses or distributed query processing) costly to develop and maintain.

As a result, there is increasing interest in approaches with reduced up-front costs (e.g. [4]), which in turn often leads to more loosely coupled models. In the life science, a particular focus has been on workflow technologies, in which services interoperate, but data need not be "integrated" in any meaningful sense. Such platforms provide consistent access to data and computational resources, and may yet provide a framework within which different data integration technologies can be brought together, accommodating as they do both pay-as-you-go [8] and plan-ahead [9] integration. However, understanding the relative costs and benefits of different data integration techniques continues to be a challenging undertaking [5], and no less so in the life sciences than elsewhere.

As such, data integration in the life sciences potentially involves both *fun* and *frustrations* while trying to produce *findings*: *fun* in that the area is a source of worthwhile problems involving diverse collaborators; and *frustrations* in that the domain continues to manifest problems that elude elegant solutions. The latter in turn means that individual projects rarely generate findings of value across the range depicted in Figure 1.

Acknowledgement. Research on data integration in the life sciences at Manchester is supported by the BBSRC and the EPSRC, whose support we are pleased to acknowledge.

References

1. Belhajjame, K., et al.: Automatic annotation of web services based on workflow definitions. In: International Semantic Web Conference, pp. 116–129 (2006)
2. Cornell, M.J., et al.: Comparative genome analysis across a kingdom of eukaryotic organisms: specialization and diversification in the fungi. Genome Research 17(12), 1809–1822 (2007)
3. Goble, C.A., et al.: Transparent access to multiple bioinformatics information sources. IBM Systems Journal 40(2), 532–551 (2001)
4. Halevy, A.Y., et al.: Principles of dataspace systems. In: PODS, pp. 1–9 (2006)
5. Howe, B., et al.: Smoothing the roi curve for scientific data management applications. In: CIDR, pp. 185–195 (2007)
6. Jones, A.R., et al.: The Functional Genomics Experiment model (FuGE): an extensible framework for standards in functional genomics. Nature Biotech. 17(12), 1809–1822 (2007)
7. King, R.D., et al.: Functional genomic hypothesis generation and experimentation by a robot scientist. Nature 427(6971), 247–252 (2004)
8. Oinn, T.M., et al.: Taverna: lessons in creating a workflow environment for the life sciences. Concurrency and Computation: Practice and Experience 18(10), 1067–1100 (2006)
9. Syed, J., et al.: Supporting scientific discovery processes in discovery net. Concurrency and Computation: Practice and Experience 19(2), 167–179 (2007)

Analyzing the Evolution of
Life Science Ontologies and Mappings

Michael Hartung[1], Toralf Kirsten[1], and Erhard Rahm[1,2]

[1] Interdisciplinary Center for Bioinformatics, University of Leipzig
[2] Dept. of Computer Science, University of Leipzig
{hartung,tkirsten}@izbi.uni-leipzig.de,
rahm@informatik.uni-leipzig.de

Abstract. Ontologies are heavily developed and used in life sciences and undergo continuous changes. However, the evolution of life science ontologies and references to them (e.g., annotations) is not well understood and has received little attention so far. We therefore propose a generic framework for analyzing both the evolution of ontologies and the evolution of ontology-related mappings, in particular annotations referring to ontologies and similarity (match) mappings between ontologies. We use our framework for an extensive comparative evaluation of evolution measures for 16 life science ontologies. Moreover, we analyze the evolution of annotation mappings and ontology mappings for the Gene Ontology.

Keywords: Ontology evolution, ontology matching, mapping evolution.

1 Introduction

Ontologies become increasingly important in life sciences. Usually, they provide a harmonized vocabulary describing and structuring a specific domain of interest, e.g., molecular functions of proteins or the anatomy of a species. The vocabulary consists of concepts, which are typically structured within trees or acyclic graphs where the concept nodes are interconnected by "*is-a*" and "*part-of*" relationships. Biological objects, such as genes and proteins, can be semantically and uniformly described or annotated by ontologies by associating them with the respective ontology concepts. For example, proteins are associated to concepts of the Gene Ontology to describe their protein functions and to specify processes they are involved in. The proliferation of ontologies has also generated interest in interrelating different ontologies by so called ontology mappings [1,2,7], e.g., to see which molecular functions are involved in which biological processes or which functions are localized on which cellular component.

Due to the rapid development of life science research we observe that ontologies evolve continuously, i.e., they are frequently changed to incorporate new domain knowledge into them. Typical ontology modifications include the addition of new concepts and new relationships or the deletion of outdated concepts and relationships. To still provide some stability for applications and users of ontologies, the ontology

A. Bairoch, S. Cohen-Boulakia, and C. Froidevaux (Eds.): DILS 2008, LNBI 5109, pp. 11–27, 2008.

developers typically support a version concept. An ontology version represents the state of the ontology at a specific point in time (release date). While older ontology versions remain stable (unchanged), a new ontology version may reflect an arbitrary number of changes. However, these changes, e.g., deletions, may impair the correctness of previous use cases of the ontology within annotations or ontology mappings. Hence, annotations and ontology mappings affected by ontology changes may have to be identified and corrected. Furthermore, new knowledge represented by added concepts and added relationships should be utilized as quickly as possible.

So far, the evolution of life science ontologies and change impact for annotations and ontology mappings has received almost no attention and is therefore not well understood. As a first step in dealing with ontology evolution in life sciences we therefore propose to analyse how existing ontologies evolve, e.g., to answer immediate questions such as "How volatile (stable) are different ontologies?" "What is the frequency of different types of modifications?" and "Which structural changes occur within ontologies?". Furthermore, we want to analyze the consequences of ontology changes, e.g., to what degree do they imply changes of ontology-based annotation and previously determined ontology mappings.

To that end, we make the following contributions in this paper:

- We propose a generic framework allowing us to systematically study the evolution of ontologies and instance data sources (e.g., representing biological objects such as proteins), as well as the evolution of ontology-related mappings, i.e., annotation mappings and ontology mappings. The framework supports the computation of several general measures to describe individual ontology versions and mappings as well as their evolution.

- In a comprehensive evaluation, we apply the framework to 386 versions of 16 life science ontologies including the sub-ontologies of Gene Ontology and the NCI (National Cancer Institute) thesaurus. In particular, we use the proposed framework measures to analyze the major change types and other evolution characteristics.

- We further evaluate the evolution of annotation mappings and correlate between changes of instances/ontologies and the ontology-based annotations. Furthermore, we analyze the impact of ontology evolution to differently generated ontology mappings.

The analysis results are expected to be helpful for both ontology developers and ontology users to better understand the consequences of ontology changes. Furthermore, the results may help guide the development of algorithms to generate mappings that remain comparatively robust against ontology changes.

The rest of the paper is organized as follows. In Section 2 we introduce a general framework to measure different types of evolutionary changes of ontologies, their associations to biological objects and on interconnecting ontology mappings. In Section 3 we apply the framework and show results for a selected set of life science ontologies whereas Section 4 illustrates the evolution results of protein objects established ontology mappings we observed. Section 5 discusses related work. We finally conclude and outline future work.

2   Evolution and Measurement Framework

Our evolution framework distinguishes between two basic types of evolution as illustrated in Figure 1. On the one side, we investigate the evolution in single *sources*, specifically *ontologies* (1) and *instance sources* (2). For both source types, the evolution is reflected in a series of versions. On the other side, we consider the evolution of *mappings*. Such mappings exist between versions of different instance sources (*instance-instance-mapping* (3)), between versions of instance sources and ontologies (*annotation mapping* (4)) and between versions of different ontologies (*ontology mapping* (5)). In the following we define the models and measures of our framework. A simple example (Figure 2) will illustrate these models and their evolution.

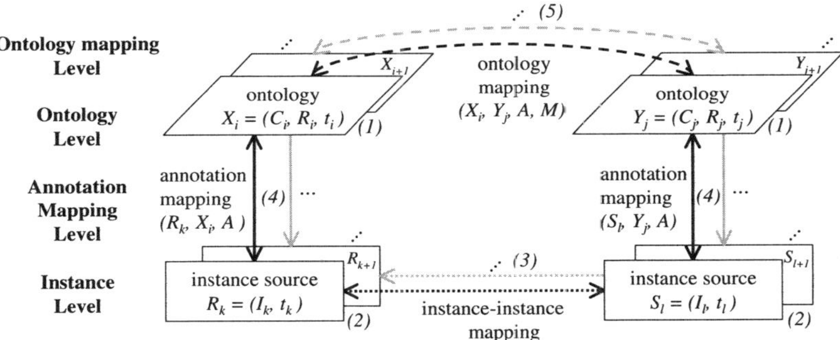

Fig. 1. Evolution of sources (1, 2) and mappings (3, 4, 5)

2.1   Framework Models

2.1.1   Ontology Model

An ontology $ON_v = (C, R, t)$ is defined by its name ON, a version number v, *concepts* $C = \{c_1, ..., c_m\}$, *relationships* $R = \{r_1, ..., r_n\}$ and a creation timestamp t. Concepts represent entities of the domain to be modeled; they are interconnected by the relationships in R, e.g., is-a and part-of relationships. Concepts with no relationships to any super concept act as the *roots* $\subseteq C$ of ON_v. Together, C, R and the *roots* form the ontology's graph structure which is assumed to be a directed acyclic graph (DAG).

A concept can have a varying number of *attributes*. Typical attributes in biomedical ontologies are accession ID, concept name, concept synonyms, concept definition, and obsolete status. In our evolution framework we heavily take into account accession ID and obsolete status information. The accession IDs unambiguously identify concepts and can be used to determine new and deleted concepts when comparing different versions of an ontology. Furthermore, these IDs are used within annotation and ontology mappings. The obsolete status is not generally supported but allows the specification of outdated concepts which may still be in use but should not be used anymore for new applications.

R defines directed binary relationships between concepts. We distinguish between three types of relationships, namely *is-a* (R_{is_a}), *part-of* (R_{part_of}) and *miscellaneous*

(R_{mis}). As we will see, is-a and part-of relationships are the most common relationship types in biomedical ontologies. Other ("miscellaneous") relationship types are specific to ontologies of a certain domain, e.g., anatomy, chemistry or molecular biology.

2.1.2 Instance Model

An instance source $IS_v = (I, t)$ of version number v consists of a set of *instances* $I = \{i_1, ..., i_n\}$, e.g., molecular biological objects such as genes or proteins, and a creation timestamp t. Instances are described by a set of attributes including an accession ID attribute and *IS*-specific attributes. The ID attribute is used in mappings between different instance sources (instance-instance mapping) and in annotation mappings.

2.1.3 Annotation Mapping Model

An annotation mapping $AM = (IS_u, ON_v, A)$ describes a mapping between an instance source *IS* of version u and an ontology *ON* of version v. The mapping itself, denoted by A, is a set of binary associations between instances I of IS_u and concepts C of ON_v. A single association or correspondence $a_j = (i_j, c_j) \in A$ annotates an instance item $i_j \in I$ with an ontology concept $c_j \in C$. Note that annotation mappings are (implicitly) versioned by the use of versioned instance sources and versioned ontologies. Hence, the combination of the version numbers u and v can be thought of as the version number of the mapping.

2.1.4 Ontology Mapping Model

We define an ontology mapping $OM = (X_u, Y_v, A, M)$ between two different ontology versions X_u and Y_v as a set of correspondences A based on a match algorithm M. A single correspondence $n_k = (x_k, y_k, sim_k) \in A$ comprises two ontology concepts (concept x_k of X_u, concept y_k of Y_v) and a similarity value sim_k. The similarity value indicates the strength of similarity between two ontology concepts and is typically a numerical value from the interval [0,1]. Similarity values are determined by an ontology match algorithm M. For example, metadata-based matching algorithms use metadata for matching such as concept names and often apply string similarity measures to estimate the similarity of ontology concepts. On the other hand, instance-based matchers may consider the number of shared instances, i.e., instances associated to both ontology concepts, to compute a similarity value [7].

Similar to annotation mappings, ontology mappings are implicitly versioned by the use of versioned ontologies.

2.1.5 Common Evolution Model

In order to analyze the evolution of single sources and of mappings, we define a generic evolution model that is applicable to all defined models, in particular ontologies, instances, annotations and ontology mappings. The basis of our evolution model are *object sets* O_{vi} of a version v_i of a source that evolves. Possible objects are ontology concepts or relationships (ontology evolution), instance data (instance evolution), annotation associations (annotation mapping evolution) and ontology correspondences (ontology mapping evolution).

We focus on three change operations that may occur during evolution: *add*, *delete* and *toObs*. Whereas *add* is used to insert new objects in a source or mapping, the *delete* operation directly removes objects which are outdated or no longer required.

ToObs is a special operation preferentially used in ontologies to mark objects as obsolete. In contrast to *delete*, obsolete objects remain in an evolved source. For simplicity and to preserve the applicability of our evolution model to both ontologies and mappings, we do not consider more complex evolution operations in this study, e.g., moves of concepts within is-a /part-of hierarchies or changes of relationship types.

To quantify the evolution behavior, for each change operation we determine the sets of affected objects in the considered source and mapping versions:

- $add_{vi,vj} = O_{vj} / O_{vi}$: *added objects* between version v_i and v_j
- $del_{vi,vj} = O_{vj} / O_{vi}$: *deleted objects* between version v_i and v_j
- $toObs_{vi,vj} = O_{vj,obs} \cap O_{vi,nonObs}$: objects that were marked as *obsolete* between version v_i and v_j. Here, the subsets $O_{vi,nonObs}$ and $O_{vi,obs}$ are used to distinguish between normal and obsolete objects in a version v_i, together they form the set of all objects O_{vi} in version v_i.

These sets can be quite easily determined for existing ontologies, instance sources, and mappings by analyzing and comparing the accession attributes of objects. For example, if an object ID is present in a newer version of a source and not in the older one, we assign this object to the *add* set, and vice versa for the *delete* set.

A simple yet comprehensive example for both ontology evolution and mapping evolution is shown in Figure 2. The example captures the evolution of two ontologies X (X_1 to X_2) and Y (Y_1 to Y_2), the evolution of one instance source I (I_1 to I_2), the evolution of two annotation mappings I-X (I_1-X_1 to I_2-X_2) and I-Y (I_1-Y_1 to I_2-Y_2), and the evolution of one ontology mapping X-Y (X_1-Y_1 to X_2-Y_2). So in ontology version X_2 there is one new concept, x_4, while concept x_3 has been declared as obsolete. For x_4, there is a new instance annotation (i_4-x_4) as well as a new ontology correspondence (x_4-y_5). For x_3, the previous instance annotation i_3-x_3 and ontology correspondence x_3-y_4 have been deleted in the new mappings.

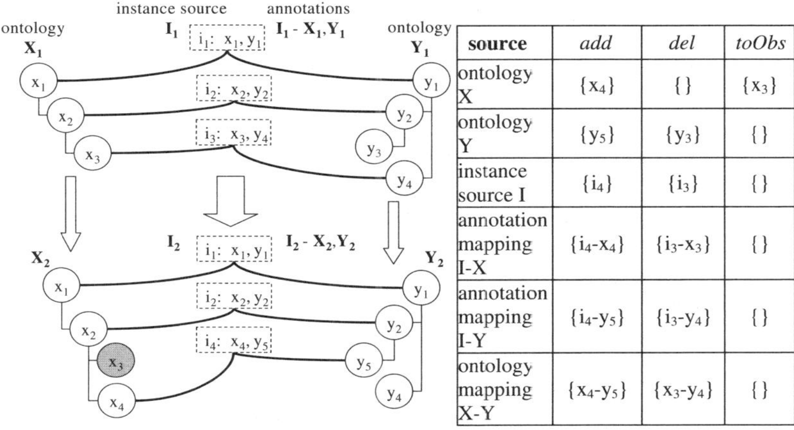

source	add	del	toObs
ontology X	{x₄}	{}	{x₃}
ontology Y	{y₅}	{y₃}	{}
instance source I	{i₄}	{i₃}	{}
annotation mapping I-X	{i₄-x₄}	{i₃-x₃}	{}
annotation mapping I-Y	{i₄-y₅}	{i₃-y₄}	{}
ontology mapping X-Y	{x₄-y₅}	{x₃-y₄}	{}

Fig. 2. Evolution example with ontologies (X,Y), instance sources (I), annotation mappings (I-X,Y) and an ontology mapping (X-Y)

2.2 Framework Measures

Based on the introduced framework, we determine a variety of statistical measures on the investigated sources (ontologies, instance sources) and mappings, as well as on their evolution and growth characteristics. We first present the source- and mapping-specific measures, followed by the evolution and growth measures.

2.2.1 Descriptive Statistics for Sources and Mappings

For all kinds of object sets (instances, concepts, relationships, correspondences), we consider their cardinality in a given version of an instance source, ontology or mapping. For ontologies, we additionally determine structural characteristics such as the used relationship types (is-a, part-of), concept types (obsolete or non-obsolete, leaf or inner concepts), in-degrees and out-degrees, as well as the number of paths and path lengths:

$|O_{vi}|$	number of *objects* in version v_i of a source or mapping $O \in$ {ontology concepts C, ontology relationships R, instance data I, annotation mapping A, ontology mapping A}
$|C_{leaf}|$, $|C_{inner}|$	number of leaf and inner concepts
$|C_{obs}|$, $|C_{nonObs}|$	number of obsolete and non obsolete concepts
$|R_{is\_a}|$, $|R_{part\_of}|$, $|R_{mis}|$	number of is-a, part-of or miscellaneous relationships
$\varnothing d_{in} = |C_{inner}| / (|R_{is\_a}|+|R_{part\_of}|)$	average in-degree of inner concepts
$\varnothing d_{out} = |C| / (|R_{is\_a}|+|R_{part\_of}|)$	average out-degree of concepts
$\varnothing ppc$, $\varnothing ppl$	average number of paths per concept or per leaf concept (path as way to a root concept using is-a or part-of relationships)
$\varnothing pl$, $\varnothing pl_{leaf}$	average path length of all concepts or leaf concepts

For mappings, let $X_{A,u} \subseteq X_u$ and $Y_{A,v} \subseteq Y_v$ be two object sets of version u and v, such that a mapping A interrelates each element of $X_{A,u}$ with at least one element of $Y_{A,v}$ and each element of $Y_{A,v}$ have at least one counterpart in $X_{A,u}$. Then, we can determine the relative coverage of X_u and Y_v for mapping A by $X_{A,u}$ and $Y_{A,v}$, respectively, i.e., the fraction of objects of X_u (Y_v) for which at least one counterpart (and thus correspondence) in mapping A exists.

$$cov_{A,Xu} = |X_{A,u}| / |X_u|$$
$$cov_{A,Yv} = |Y_{A,v}| / |Y_v|$$

relative coverage of objects X_u and Y_v by the mapping A

2.2.2 Evolution and Growth Statistics

Our measures make use of the generic evolution model to compute evolution statistics for all evolution types (ontologies, instance sources, mappings). To determine the number of changes or changed objects we either directly compare two versions v_i and v_j of a source or mapping. Alternatively, we quantify the changes with respect to a certain time interval, e.g., for an entire observation period p or a regular time interval t within p, e.g., per month or per year.

$Add_{vi,vj} = |add_{vi,vj}|$	number of *added objects* between version v_i and v_j
$Del_{vi,vj} = |del_{vi,vj}|$	number of *deleted objects* between version v_i and v_j
$Obs_{vi,vj} = |toObs_{vi,vj}|$	number of objects that changed to obsolete between version v_i and v_j
$Add_{p,t}$ $Del_{p,t}$ $Obs_{p,t}$	average number of *added / deleted / obsolete* objects per time interval t within p

Based on these basic frequencies we determine relative fractions of newly added and deleted objects as well as an *add-delete ratio* (adr) between two versions. Further, we quantify relative fractions relating to a certain time interval t within a period p:

$$adr_{vi,vj} = Add_{vi,vj} / (Del_{vi,vj} + Obs_{vi,vj})$$

add-delete ratio for changes between version v_i and v_j

$$add-frac_{vi,vj} = \frac{Add_{vi,vj}}{|O_{vj}|}$$

fraction of objects in version v_j that have been added between version v_i and v_j

$$del-frac_{vj,vi} = \frac{Del_{vi,vj}}{|O_{vi}|}$$

fraction of objects in version v_i that have been deleted between version v_i and v_j

$$obs-frac_{vj,vi} = \frac{Obs_{vi,vj}}{|O_{vi}|}$$

fraction of objects in version v_i that have been marked as obsolete between version v_i and v_j

$$add-frac_{p,t} \quad del-frac_{p,t} \quad obs-frac_{p,t}$$

average fractions of *added* / *deleted* / *obsolete* objects per time interval t within p based on the version-related *frac* measures

We further define *growth rates*

$$growth_{O,vi,vj} = |O_{vj}| / |O_{vi}| \in [0, \infty] \subseteq R$$

for most of the measures above as the ratio between the objects O ($O \in$ {ontology concepts C, ontology relationships R, instance data I, annotation mapping A, ontology mapping A}) of version v_j and v_i. The growth rate describes an increase when the rate is greater than 1, a decrease when the rate is less than 1 or no change for $growth_{vi,vj}=1$. Moreover, the growth rate can also be determined for relative measures, such as fractions or coverages, e.g., an increase from 50% to 60% for the ontology coverage between two versions of an ontology mapping corresponds to a growth rate of 1.2.

3 Analysis of Ontology Evolution

We study the evolution of ontologies of different life science domains, ranging from popular Gene Ontology (GO) [3] and NCI Thesaurus [12] to more specific ontologies of the OBO foundry [16], e.g., SequenceOntology or ZebrafishAnatomy. In order to comparatively analyze these ontologies, we set up a central repository with a generic schema suitable for management of heterogeneous ontologies and their versions. Overall, we integrated 386 versions of 16 currently developed life science ontologies.

In the following, we first give an overview of the analyzed ontologies and their versions. We then use the introduced measures to study the evolution behavior of the ontologies including structural ontology changes. Exemplary evolution trend charts for GO Biological Processes and Molecular Functions will be presented in Section 4.2. Detailed information and evolution trend charts for all analyzed ontologies can be found in [5] and online (http://dbs.uni-leipzig.de/ls_ontology_evolution).

3.1 Overview and Versioning Aspects

Table 1 lists the ontologies and gives details about their size, the number of versions during the observation period, the growth ratio as well as domain and use characteristics. For clarity, we group the analyzed ontologies into 3 groups (*large, medium, small*) based on their current number of concepts |C|. Our evaluation considers ontology versions for an observation period of 45 months, from May 2004 until Feb. 2008. The timestamps t_{start} (t_{last}) of the first (latest) version and the number of versions (k)

Table 1. Overview and versioning statistics of analyzed ontologies. *Size categories* - small: $|C|$ < *1000*, medium: *1000 < |C| < 10000*, large: $|C| > 10000$

Ontology	size	$|C|_{start}$	$|C|_{last}$	$grow_{|C|, start, last}$	t_{start}	t_{last}	k	characteristics, domain and use
NCI Thesaurus		35,814	63,924	1.78	May. 04	Dec. 07	39	broad coverage of cancer domain
GeneOntology		17,368	25,995	1.50	May. 04	Feb. 08	44	aggregation of all GO sub ontologies
– Biological Process		8,625	15,001	1.74	May. 04	Feb. 08	44	annotation of gene products (biological role)
– Molecular Function	large	7,336	8,818	1.20	May. 04	Feb. 08	44	annotation of gene products (molecular function)
– Cellular Components		1,407	2,176	1.55	May. 04	Feb. 08	44	annotation of gene products (cellular location)
ChemicalEntities		10,236	18,007	1.76	Oct. 04	Jan. 08	28	chemical compounds of biological relevance
FlyAnatomy		6,090	6,222	1.02	Nov. 04	Dec. 07	16	anatomy of Drosophila melanogaster
MammalianPhenotype		4,175	6,077	1.46	Aug. 05	Jan. 08	15	terms for annotating mammalian phenotypic data
AdultMouseAnatomy	medium	2,416	2,745	1.14	Aug. 05	Sep. 07	15	adult anatomy of the mouse (Mus)
ZebrafishAnatomy		1,389	2,172	1.56	Nov. 05	Oct. 07	12	anatomy and development of the Zebrafish
Sequence		981	1,463	1.49	Aug. 05	Feb. 08	26	structured CV for sequence annotation
ProteinModification		1,074	1,128	1.05	Jun. 06	Nov. 07	14	description of protein chemical modifications
CellType		687	857	1.25	Jun. 04	Jun. 07	19	cell types from prokaryotes to mammals
PlantStructure		681	835	1.23	Jul. 05	Feb. 08	22	plant morphological and anatomical structures
ProteinProteinInteraction	small	194	819	4.22	Aug. 05	Feb. 08	19	annotation of protein interaction experiments
FlyBaseCV		658	693	1.05	Nov. 05	Apr. 07	7	used for various aspects of annotation by FlyBase
Pathway		427	593	1.39	Nov. 05	Jan. 08	22	CV for pathways, annotation of gene products
Overall		82,190	131,530	1.60			386	

provide information about the versioning rate of an ontology, i.e., how often an ontology releases versions and how long they are actively used. While some ontologies, particularly the Gene Ontology, currently release versions every day we consider at most one version per month (for several versions per month, we pick the first one). We observe that the oldest and most frequently released ontologies are the two largest ontologies, NCI Thesaurus and Gene Ontology. Other ontologies such as FlyBaseCV or CellType have not been updated since a longer period (6-8 months) which may indicate that these ontologies have reached a near-final state. The average number of versions per ontology is 25, i.e., a version is typically current for less than 2 months.

In terms of number of concepts, we observe a considerable growth during the observation period. On average, the number of concepts has increased by 60% during the last 45 months; the maximum (minimum) growth rate is 4.22 (1.02). The largest ontology, NCI Thesaurus has increased its size by 80% to almost 64,000 concepts. The largest and fastest growing GO subontology is Biological Processes (74% increase); on the other hand, the number of Molecular Functions concepts has merely increased by 20% during the observation period.

Table 2 shows more detailed and time-normalized statistics on the evolution behavior of the considered ontologies. In particular, it indicates the average number of newly added, deleted and obsolete concepts *per month* for both the entire observation period and the last year only. In addition, the relative fractions of concepts are specified which are added, deleted or declared obsolete per month.

We observe that the largest ontologies experience the highest numbers in changes. On average, they have approx. 360 (25) additions (deletions) per month compared to approx. 86 (6) additions (deletions) in all analyzed ontologies. Furthermore, the study shows that additions are the dominant change operation for all ontologies. Still, some ontologies experience a significant number of deletions, e.g., ChemicalEntities and Gene Ontology. The add-delete ratio (*adr*) indicates the relative frequency of these two main change types. NCI Thesaurus has the maximal value of 42, indicating that there are 42 times more additions than deletions or new obsolete cases. On the other

Table 2. Evolution of analyzed life science ontologies (interval $t = 1$ month)

Ontology	Full period (May. 04 - Feb. 08)							Last year (Feb. 07 - Feb. 08)		
	Add	Del	Obs	adr	add-frac	del-frac	obs-frac	Add	Del	Obs
NCI Thesaurus	627	2	12	42.4	1.3%	0.0%	0.0%	416	0	5
GeneOntology	200	12	4	12.2	0.9%	0.1%	0.0%	222	20	5
-- Biological Process	146	7	2	16.2	1.2%	0.1%	0.0%	133	10	2
-- Molecular Function	36	3	2	6.8	0.4%	0.0%	0.0%	69	7	3
-- Cellular Components	18	2	0	8.9	1.0%	0.1%	0.0%	19	3	0
ChemicalEntities	256	62	0	4.1	1.8%	0.5%	0.0%	384	67	0
FlyAnatomy	5	1	1	3.3	0.1%	0.0%	0.0%	6	0	0
MammalianPhenotype	65	2	9	6.0	1.2%	0.0%	0.2%	74	2	3
AdultMouseAnatomy	11	0	0	30.9	0.4%	0.0%	0.0%	1	0	0
ZebrafishAnatomy	33	5	1	5.5	1.8%	0.3%	0.1%	45	2	1
Sequence	19	3	2	4.1	1.5%	0.3%	0.2%	19	0	0
ProteinModification	5	2	1	1.5	0.4%	0.2%	0.1%	7	0	2
CellType	5	1	0	2.8	0.7%	0.2%	0.1%	1	0	0
PlantStructure	5	0	1	6.1	0.7%	0.0%	0.1%	3	0	0
ProteinProteinInteraction	21	0	0	41.7	2.7%	0.0%	0.2%	4	0	0
FlyBaseCV	1	0	1	2.1	0.2%	0.0%	0.1%	0	0	0
Pathway	7	1	0	7.9	1.3%	0.2%	0.0%	6	2	0

hand, for ChemicalEntities this ratio is merely 4, i.e., about 20% of the changes are deletes. The relative change fractions reveal that some small and medium ontologies have high evolution rates. In terms of additions, ProteinProteinInteraction has the highest relative change frequency (2.7% new concepts per month).

Another interesting observation is the usage of the obsolete paradigm in different ontologies. Some ontology designers do not mark outdated ontology concepts as obsolete, but strictly delete them, e.g., ChemicalEntities or AdultMouseAnatomy. Most ontologies (13 of 16) follow a hybrid approach, i.e., they use both to-obsolete and delete operations. Some ontologies (NCI Thesaurus, MammalianPhenotypes), perform few deletes but primarily use the obsolete status to mark outdated concepts.

Comparing the evolution rates of the last year with the ones of the overall observation period allows us to see recent evolution trends for the different ontologies. A first group of ontologies exhibits high evolution rates in both periods, e.g., NCI Thesaurus, GO or MammalianPhenotype. This indicates that the knowledge in the domains of these ontologies is continuously evolving and that these ontologies refer to active research fields. A second group of ontologies has considerably higher evolution rates in the last year indicating an increased research activity in the respective domains, e.g., for ChemicalEntities or GO Molecular Function. Finally, we discover ontologies with few changes in the recent past, e.g., AdultMouseAnatomy, CellType or FlyBaseCV. Work on these ontologies may have almost been finished so that rather stable ontology versions can be used.

3.2   Influence of Evolution on Ontology Structures

Due to space limitations, we analyze the evolution of structural properties only for the largest ontologies. Table 3 summarizes structural measures for the first and last version of the considered 6 ontologies as well as the resulting growth rates (lower third of the table). We consider the evolution in the relative share of leaf (vs. inner) nodes, the number of relationships, the distribution of is-a, part-of and other relationships, as well as in the concept node degrees and number of paths.

Table 3. Changes in ontology structures

	Ontology	$|C_{leaf}|$ (%)	$|R|$	$|R_{isa}|$ (%)	$|R_{partof}|$ (%)	$|R_{mis}|$ (%)	$\varnothing\, d_{out}$	$\varnothing\, d_{in}$	$\varnothing\, pl_{leaf}$	$\varnothing\, ppl$
First version	NCI Thesaurus	79	41,281	100			1.2	5.6	8.2	3.3
	GeneOntology	66	23,589	88	12		1.4	4.0	7.3	3.7
	-- Biological Process	52	13,358	85	15		1.5	3.2	8.0	7.1
	-- Molecular Function	82	8,459	100			1.2	6.4	5.3	1.4
	-- Cellular Components	67	1,772	52	48		1.3	3.8	5.5	1.8
	ChemicalEntities	70	11,593	100			1.1	3.8	8.3	2.3
	MammalianPhenotype	68	4,620	100			1.1	3.4	5.5	1.5
Last version	NCI Thesaurus	79	72,466	100			1.1	5.4	8.0	3.0
	GeneOntology	60	41,396	88	12		1.6	3.8	8.6	22.9
	-- Biological Process	46	27,141	84	16		1.8	3.3	8.8	38.7
	-- Molecular Function	81	10,195	100			1.2	5.9	6.2	1.7
	-- Cellular Components	64	4,060	79	21		1.9	5.0	8.3	52.6
	ChemicalEntities	69	31,233	76	1	23	1.4	4.3	12	18.6
	MammalianPhenotype	64	6,875	100			1.2	3.1	7.5	2.5
Growth	NCI Thesaurus	1.00	1.8	1.00			1.0	1.0	1.0	0.9
	GeneOntology	0.91	1.8	1.00	1.03		1.2	1.0	1.2	6.2
	-- Biological Process	0.89	2.0	0.99	1.06		1.2	1.0	1.1	5.5
	-- Molecular Function	1.00	1.2	1.00			1.0	0.9	1.2	1.2
	-- Cellular Components	0.95	2.3	1.51	0.44		1.5	1.3	1.5	28.7
	ChemicalEntities	0.99	2.7	0.76	undef.	undef.	1.3	1.1	1.4	8.0
	MammalianPhenotype	0.95	1.5	1.00			1.1	0.9	1.4	1.7

We observe that for the considered ontologies, the majority of concepts is represented by leaf nodes, i.e., these concepts are not further refined by is-a or part-of relationships. However, the relative share of leaf nodes has reduced during the observation period from about 70% to 67% indicating a corresponding increase of inner nodes and in structured knowledge. For one ontology, GO Biological Process, there are now even fewer leaf concepts (46%) than inner concepts due to a strong decline in the fraction of leaf nodes ("growth" rate 0.89).

The number of relationships increased similarly or faster than the number of concepts (Table 1) during the observation period. The largest increase occurred for ChemicalEntities (growth factor 2.7 for relationships vs. 1.76 for concepts). The considered ontologies are dominated by is-a relationships (ca. 91% of all relationships), while part-of (4%) and miscellaneous (5%) relationships are similarly infrequent[1]. Some ontologies are pure is-a hierarchies, e.g., NCI Thesaurus, GO Molecular Function or MammalianPhenotype. The biggest changes occurred for ChemicalEntities which started as a pure is-a ontology but introduced part-of and other relationship types in recent versions. We also observe interesting differences between the GO subontologies. While Molecular Function only uses is-a relationships, Biological Process and Cellular Components contain both is-a and part-of relationships. However, the relative share of part-of evolved differently: Biological Process now relies more on part-of than in the beginning (growth: 1.06) while Cellular Components has a sharp relative reduction for part-of (0.44).

With respect to the in-degrees and out-degrees of concept nodes we notice little changes during the observation period, especially for is-a ontologies. The out-degrees of these ontologies is typically lower than 1.2, i.e., their concepts have mostly only

---

[1] With respect to all 16 ontologies, the relative shares for is-a / part-of / miscellaneous relationships are 86% / 7% / 7%.

one super concept. On the other side, ontologies such as GO Cellular Components or GO Biological Process have about two ancestor concepts per concept since they use is-a and part-of relationships in combination. Lastly, we look at the evolution of path lengths and number of paths in leaf concepts. We notice that except NCI Thesaurus all ontologies increased in their average path length of leaf concepts (up to 50%). The number of paths per leaf ($\varnothing\,ppl$) heavily increased, especially for ontologies which are not limited to is-a relationships (average growth rate: 14). The highest growth rate (28) occurred for the GO Cellular Components which apparently experienced a major restructuring as already observed for the development of is-a vs. part-of relationships.

4 Evolution of Annotation and Ontology Mappings

In further experiments we studied the evolution of the annotation and ontology mappings. We start with a short overview of the scenario we used in the evaluation before we describe the obtained results.

4.1 Evaluation Scenario

Figure 3 shows a schematic overview of the evaluation scenario. To reduce the complexity we focus on two ontologies, namely the GO subontologies Molecular Functions and Biological Processes. Both ontologies are usually used to describe properties of proteins, i.e., the function and process concepts of the ontologies are associated with proteins. We therefore evaluate protein instances, namely protein objects of the human species available in the data source Ensembl [6]. Furthermore, we analyze the annotation mappings, as provided by Ensembl, between these proteins and the two ontologies. To interrelate the two ontologies, we determine different ontology mappings using either metadata-based or instance-based match algorithms. We will give some more details below.

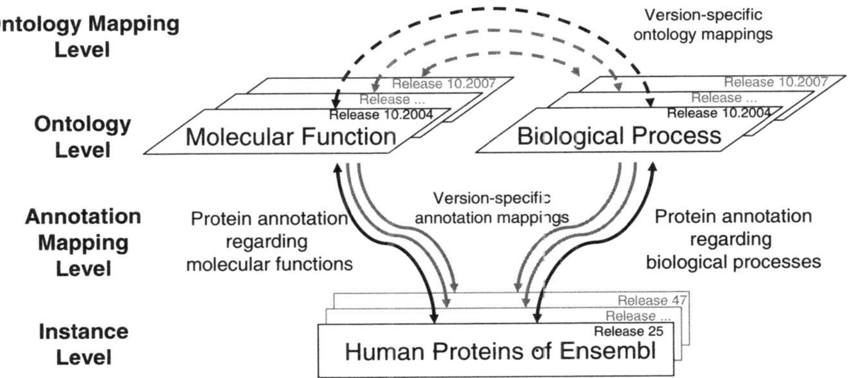

Fig. 3. Overview about the evaluation scenario

22   M. Hartung, T. Kirsten, and E. Rahm

4.2 Evolution of Instance Source vs. Ontologies

The Ensembl instances and annotations as well as the two ontologies underlie frequent changes. The evaluation scenario includes 23 versions of Ensembl from Oct. 2004 to Oct. 2007 (36 months). Table 4 shows the Ensembl release numbers together with their release month and year. While in 2004 and 2005 the Ensembl releases appeared irregularly, since 2006 a new Ensembl release is created every two months. The Ensembl information is heavily based on the genome assemblies made public by NCBI; since 2004, three such assemblies (namely 34, 35, and 36) have appeared. Moreover, Table 4 also shows which GO releases have been used for the annotation mappings provided in Ensembl. As one can see, the annotations typically do not refer to the most recent but an older GO version. For example, the annotation mapping in Ensembl release 37 (Feb. 2006) refers to the GO version of March 2005, i.e., there is a time delay of 11 months. The delay has been reduced in recent Ensembl versions.

Figure 4a illustrates the evolution of protein objects (total number, number of added and deleted instances) in Ensembl from Oct. 2004 to Oct. 2007. We observe that a new genome assembly (Nov. 2004, Apr. 2006) led to massive changes of protein objects. The change from version 34 to 35 of the genome assembly caused many protein additions and deletions while the total number of proteins remained almost unchanged. However, the change from version 35 to 36 (April 2006), resulted in five

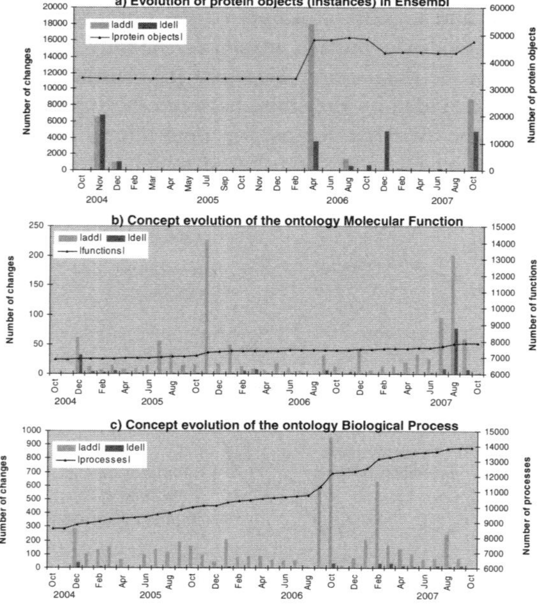

Fig. 4. Evolution of instance data and ontologies

Table 4. Release states of protein objects in Ensembl

Year	Time	Ensembl Release	NCBI Genome	Used GO Release
2004	Oct.	25	34	02.2004
2004	Nov.	26	35	02.2004
2004	Dec.	27	35	09.2004
2005	Feb.	28	35	09.2004
2005	Mar.	29	35	09.2004
2005	Apr	30	35	09.2004
2005	May	31	35	03.2005
2005	July	32	35	03.2005
2005	Sep.	33	35	03.2005
2005	Oct.	34	35	03.2005
2005	Nov.	35	35	03.2005
2005	Dec.	36	35	03.2005
2006	Feb.	37	36	03.2006
2006	Apr.	38	36	03.2006
2006	June	39	36	03.2006
2006	Aug.	40	36	03.2006
2006	Oct.	41	36	03.2006
2006	Dec.	42	36	09.2006
2007	Feb.	43	36	09.2006
2007	Apr.	44	36	03.2007
2007	June	45	36	05.2007
2007	Aug.	46	36	06.2007
2007	Oct.	47	36	06.2007

times more added than deleted proteins and a corresponding jump in the total number of proteins (about 14,000 more proteins). Recently, there are more changes on protein objects during the utilization of genome assembly 36.

For comparison, Figures 4b and 4c show the evolution of the two considered ontologies during the observation period since Oct. 2004. In contrast to the irregular evolution pattern of Ensembl, we observe that both ontologies experience a continuous evolution with added and deleted/to-obsolete concepts, despite the existence of several peaks in the number of changes. With respect to the growth in the number of objects, the Molecular Function (MF) ontology evolved the least (growth 1.2) and slower than the number of protein instances (growth 1.39 for the entire observation period). The fastest growth is observed for the Biological Processes (BP) ontology (1.74). Furthermore, there are primarily additions and few deletes for the ontologies (add-delete ratios of about 7 and 16 for MF and BP, respectively) while there is significant delete activity for the protein instances (add-delete ratio of 1.6).

4.3 Evolution on Annotation Mapping Level

In this analysis we focus on the evolution of the two Ensembl annotation mappings proteins-MF and proteins-BP. For both cases, we compare two versions namely the annotation mappings of Ensembl release 25 (Oct. 2004, first in this study) with those of release 47 (Oct. 2007, last in this study). Table 5 shows the corresponding evolution measures, introduced in Section 2, in particular growth rates for the number of correspondences, proteins and ontology concepts as well as the add and delete fractions (Table 5a). Table 5b shows coverage measures indicating which shares of the protein source and ontologies participate in the annotation mappings and how these shares changed (growth rates) between the two Ensembl versions.

Table 5. Evolution of annotation mappings between Ensembl releases 25 and 47

Annotation Mappings	Corresp. growth		Protein obj. growth		Concepts growth	
	add-frac	del-frac	add-frac	del-frac	add-frac	del-frac
Protein-MF	2.82		1.99		1.39	
	83%	51%	68%	37%	32%	6%
Protein-BP	2.47		1.90		2.25	
	81%	52%	68%	39%	58%	5%

a) Growth rates of annotation mappings

Annotation Mappings	Protein obj.		Concepts	
	cov_{25}	cov_{47}	cov_{25}	cov_{47}
	$growth_{cov}$		$growth_{cov}$	
Protein-MF	47%	67%	28%	35%
	1.43		1.22	
Protein-BP	43%	59%	20%	26%
	1.36		1.39	

b) Coverage statistics

We observe both annotation mappings show a rather similar evolution behaviour. For both mappings, the growth rates for the total number of correspondences (annotation associations) of 2.82 and 2.47 are very high. These rates are not only higher than the growth for the total number of proteins or ontology concepts (factors between 1.2 and 1.74, see above) but also higher than for the number of annotated proteins (growth factor 1.9 – 1.99) and used ontology concepts (1.39 – 2.25). Similarly, the add and delete activity is much higher for the correspondences than for the individual sources. So the latest annotation mappings of Ensembl release 47 contain 81-83% added (i.e., new) correspondences compared to the initial mapping versions of release 25. Further, more than 50% of the original correspondences have been deleted. These observations reveal that the use of ontologies in annotations grows faster than the ontologies and the number of instances but that there is also a high degree of instability due to many deletions of associations.

This is also confirmed by the coverage ratios shown in Table 5b. The much increased number of correspondences led to an increased annotation coverage for proteins. The coverage values increased during the observation period from 43-47% to 59-67%, i.e., most proteins are now annotated with concepts of the Gene Ontology. Similarly, the coverage of the two ontologies within the annotation mappings improved. Currently, 35% (26%) of the MF (BP) concepts have associated proteins.

4.4 Evolution on Ontology Mapping Level

On the ontology mapping level, we study the evolution of mappings between different versions of the MF and BP ontologies. Such semantic mappings are to specify which molecular functions are involved in which biological processes. The manual creation of such mappings is very time-consuming especially since the ontologies change so frequently. Hence we aim at a (semi-) automatic generation of mappings by using different match algorithms to generate likely correspondences between two ontology versions. For our study we consider four match algorithms of [7]. The first two match approaches are instance-based and assume that two concepts are related if they share a certain number of instances, i.e., associated protein objects in our scenario. The approach termed *Base(5)* assumes that two concepts match if there are at least 5 proteins which associate to both concepts. The *Min(1.0)* approach uses the so-called min similarity and threshold 1.0, i.e., two concepts match if all instances associated to the concept with the smaller number of associations are also associated to the other concept. The two other match approaches are metadata-based and utilize the similarity of concept names. We assume a correspondence between concepts when the string (trigram) similarity of their names exceeds a certain threshold, e.g., 0.5 or 0.7; these mappings are named with *Name(0.5)* and *Name(0.7)*. With these approaches we generated MF-BP mappings for the ontology versions of Feb. 2004 (associated with Ensembl release 25) and June 2007 (associated with Ensembl release 47).

Table 6a shows the growth rates for the ontology mappings between the two releases as well as the relative fractions for add and delete. In the column "Corresp." we also indicate the absolute number of correspondences in the two versions of the mappings (e.g., for Base(5) the number of correspondences increased from 2780 in the old version to 8973 in the new version of the ontology mapping, growth factor 3.2). Table 6b shows the coverage rates for both ontologies and both mapping versions indicating to what degree the ontologies participate in the mappings. For example, for Base(5) the coverage of MF increased from 7% to 12% between the two versions.

We observe that there are significant differences between the mappings generated by the different match algorithms and their evolution behaviour. For the name matchers, the number of correspondences is heavily dependent on the chosen threshold. A low threshold (0.5) matches many concepts (many correspondences) and leads a relatively high coverage in the ontologies, however with the risk of many false correspondences. A higher threshold (0.7), on the other hand, is very restrictive and matches only few concepts. On the other hand, this restrictive approach leads to the highest evolution stability with the lowest fraction for deleted correspondences (17%).

Table 6. Evolution of generated ontology mappings between molecular functions and biological processes of the GeneOntology source

a) Growth rates of ontology mappings

Ontology Mappings	Corresp. $|C1|-|C2|$, grow		Mol. Functions grow		Biol. Processes grow	
	add-frac	del-frac	add-frac	del-frac	add-frac	del-frac
Base(5)	2780-8973, 3.2		1.8		2.3	
	78%	29%	52%	16%	62%	12%
Min (1.0)	4795-11564, 2.4		1.4		2.1	
	80%	52%	41%	15%	62%	21%
Name (0.5)	5434-15016, 2.8		2.1		1.4	
	77%	36%	57%	10%	44%	20%
Name (0.7)	389-592, 1.5		1.3		1.3	
	45%	17%	32%	12%	34%	15%

b) Coverage statistics

Ontology Mappings	Mol. Functions		Biol. Processes	
	cov_{25}	cov_{47}	cov_{25}	cov_{47}
	$grow_{cov}$		$grow_{cov}$	
Base(5)	7%	12%	6%	8%
	1.7		1.3	
Min (1.0)	23%	30%	17%	20%
	1.3		1.2	
Name (0.5)	25%	47%	18%	15%
	1.9		0.8	
Name (0.7)	5%	6%	4%	3%
	1.2		0.7	

Interestingly, for the name matchers the coverage of the BF ontology decreased, presumably because the BF ontology growed much faster than the MF ontology so that for many new BF names there no MF counterpart is found.

The two instance-based matchers obtain a relatively high number of correspondences (compared to the name matchers) as well as a large increase between the two versions (growth factor 2.4 – 3.2), i.e., the mappings grow faster than the ontologies. The Base(5) matcher is more stable than the Min(1.0) matcher since the delete fraction is merely 29% vs. 52%. On the other hand the Min matcher achieves a much better coverage.

5 Related Work

The evolution and change management of ontologies has so far primarily been addressed in the context of the Semantic Web [18], especially for specific ontology representations such as OWL or Frame Logic. Klein [8,9] investigated the versioning of ontologies, [10] defined change operations to describe the evolution between ontology versions. In [13,14,15], the process of ontology evolution has been formalized and strategies to unambiguously handle critical changes during evolution are proposed. Tools supporting change management of different ontology models are described in [4,11,13].

This line of previous work is complementary to ours and does neither consider life science ontologies nor a quantitative analysis of the evolution behavior. Furthermore, the evolution of ontology-related mappings has not been analyzed before. One recent paper analyzed the evolution of the Gene Ontology [17] using a simple evolution model. We also used some of their measures (e.g., number of paths or path lengths of concept nodes) but propose a more powerful generic evolution model that is applicable to the evolution of ontologies, instance sources, and mappings. Furthermore, we comparatively analyzed not only the Gene Ontology but 16 biomedical ontologies as well as the evolution of annotation and ontology mappings.

6 Conclusions

We proposed a general framework for analyzing the evolution of ontologies and ontology-related mappings. Using the framework we analyzed the recent evolution of 16 life-science ontologies since 2004. We observed that most ontologies are heavily updated and grow significantly. Most changes are additions of new concepts but there is also a surprisingly high number of concepts that are deleted in newer versions or marked as obsolete. The notion of obsolete concepts is supported by most but not all ontologies. This notion is helpful for the stability of ontologies and eases applications the migration to newer ontology versions (without risking invalid references to deleted concepts). The analyzed ontologies are dominated by is-a relationships (>85% of all relationships), although the shares of part-of and domain-specific relationships have slightly increased in recent years. Furthermore, the inner structure of ontologies (share of inner concepts, number of paths, path lengths) increased in the recent past indicating a growth of structured knowledge in life science ontologies.

We further utilized the framework to study the evolution of protein instances, annotation mappings and ontology mappings. Using Ensembl, we observed a large increase in the number of protein annotations to the Gene Ontology (GO). However, the relatively high number of deletes of protein instances caused a rather high instability for the annotation mappings. For the evolution of ontology mappings, we considered several instance- and metadata-based match algorithms to automatically generate correspondences between concepts of two GO subontologies. We observed that the ontology mappings evolved to a larger degree than the ontologies especially for the instance-based methods. Metadata-based methods (e.g., based on concept names) can easily introduce wrong correspondences but may provide improved stability for evolution. This is because they are not dependent on instances and their annotations and thus do not suffer from the higher fluctuation (delete activity) for instances compared to ontologies.

We see several opportunities for future work. First, our analysis framework can be extended by additional types of change (e.g., modification of attribute values) and applied to further ontologies. Second, algorithms to generate annotation and ontology mappings can be extended or refined to improve their stability w.r.t. ontology evolution, e.g., by taking obsolete concepts and versioning explicitly into account. Third, tools can be developed to help ontology designers to explore the effects of certain ontology changes on existing annotation and ontology mappings, especially for delete operations.

Acknowledgements. This work is supported by BMBF grant 01AK803E "MediGRID - Networked Computing Resources For Biomedical Research".

References

[1] Bodenreider, O., Aubry, M., Bugrun, A.: Non-lexical approaches to identifying associative relations in the Gene Ontology. In: Proc. Pacific Symposium on Biocomputing (2005)

[2] Bodenreider, O., Bugrun, A.: Linking the Gene Ontology to other biological ontologies. In: Proc. ISMB meeting on Bio-Ontologies (2005)

[3] The Gene Ontology Consortium: The Gene Ontology (GO) database and informatics resource. Nucleic Acids Research 32, D258–D261 (2004)

[4] Haase, P., van Harmelen, F., Huang, Z., et al.: A framework for handling inconsistency in changing ontologies. In: Proc. of 4th Intl. Semantic Web Conference (2005)

[5] Hartung, M., Kirsten, T., Rahm, E.: Analyzing the Evolution of Life Science Ontologies and Mappings - Extended Version. Leipzig Bioinformatics Working Paper No. 17 (2008)

[6] Hubbard, T., Aken, B., Beal, K., et al.: Ensembl 2007. Nucleic Acids Research 35, D610–D617 (2006)

[7] Kirsten, T., Thor, A., Rahm, E.: Instance-based matching of large life science ontologies. In: Cohen-Boulakia, S., Tannen, V. (eds.) DILS 2007. LNCS (LNBI), vol. 4544, pp. 172–187. Springer, Heidelberg (2007)

[8] Klein, M., Fensel, D.: Ontology versioning on the Semantic Web. In: Proc. Int. Semantic Web Working Symposium (SWWS) (2001)

[9] Klein, M.: Change Management for Distributed Ontologies. PhD thesis, Vrije Universiteit Amsterdam (2004)

[10] Noy, N., Klein, M.: Ontology evolution: Not the same as schema evolution. Knowledge and Information Systems 6(4), 428–440 (2004)

[11] Noy, N., Chugh, A., Liu, W., et al.: A Framework for Ontology Evolution in Collaborative Environments. In: Proc. of the 5th Intl. Semantic Web Conference (2006)

[12] Sioutos, N., de Coronado, S., Haber, M.W., et al.: NCI Thesaurus: A semantic model integrating cancer-related clinical and molecular information. Journal of Biomedical Informatics 40, 30–43 (2007)

[13] Stojanovic, L.: Methods and Tools for Ontology Evolution. PhD thesis, University of Karlsruhe (2004)

[14] Stojanovic, L., Maedche, A., Motik, B., et al.: User-driven ontology evolution management. In: Proc. of 13th Intl. Conf. On Knowledge Engineering and Knowledge management (2002)

[15] Stojanovic, L., Motik, B.: Ontology evolution within ontology editors. In: Proc. of the OntoWeb-SIG3 Workshop (2002)

[16] Smith, B., Ashburner, M., Rosse, C., et al.: The OBO Foundry: coordinated evolution of ontologies to support biomedical data integration. Nature Biotechnology 25, 1251–1255

[17] Yang, Z., Zhang, D., Ye, C.: Ontology Analysis on Complexity and Evolution Based on Conceptual Model. In: Leser, U., Naumann, F., Eckman, B. (eds.) DILS 2006. LNCS (LNBI), vol. 4075, pp. 216–223. Springer, Heidelberg (2006)

[18] Yildiz, B.: Ontology Evolution and Versioning. Technical Report, TU Vienna (2006)

Ontology Design Principles and Normalization Techniques in the Web

Xiaoshu Wang[1], Jonas S. Almeida[1,2], and Arlindo L. Oliveira[1]

[1] INESC-ID: Instituto de Engenharia de Sistemas e Computadores Investigação e
Desenvolvimento em Lisboa, R. Alves Redol 9, 1000-029 Lisboa, Portugal
[2] Department of Biostatistics and Applied Mathematics, The University of Texas M.D.
Anderson Cancer Center, 515 Holcombe Blvd, Box 0447, Houston, TX 77030
`xiao@kdbio.inesc-id.pt`, `jalmeida@mdanderson.org`,
`aml@inesc-id.pt`

Abstract. The fundamental issue of knowledge sharing in the web is the ability to share the ontological constrains associated with the Uniform Resource Identifiers (URI). To maximize the expressiveness and robustness of an ontological system in the web, each ontology should be ideally designed for a confined conceptual domain and deployed with minimal dependencies upon others. Through a retrospective analysis of the existing design of BioPAX ontologies, we illustrate the often encountered problems in ontology design and deployment. In this paper, we identify three design principles – minimal ontological commitment, granularity separation, and orthogonal domain – and two deployment techniques – intensionally normalized form (INF) and extensionally normalized form (ENF) – as the potential remedies for these problems.

Keywords: Semantic Web, Resource Description Framework (RDF), Ontology Web Language (OWL), Ontology, Uniform Resource Identifier (URI), BioPAX.

1 Introduction

As the word *ontology* is variously used, we should define our use first. *Ontology* is here defined to be the RDF graph retrieved from the Uniform Resource Identifier (URI) of an ontological term. Although the definition may seem overly restrictive, for it has limited both the ontology's formalism (to RDF) and its application (to the web), it is necessary to establish the context of this discussion: of all techniques related to ontology development, what is described in this article is applicable mostly, if not exclusively, to the ontologies deployed in the web.

As an RDF graph can be decomposed into a set of RDF triples (subject property object), an ontology is syntactically equivalent to a set of formal assertions, possibly made in a logic language, such as Ontology Web Language (OWL), and intended for modeling a particular aspect of reality. Common logics, however, concerns only the validity of inference among set of symbols but not the validity of what symbols represent. A logic language is, therefore, semantically neutral with regard to the

A. Bairoch, S. Cohen-Boulakia, and C. Froidevaux (Eds.): DILS 2008, LNBI 5109, pp. 28–43, 2008.

knowledge of its modeled domain. To make itself useful in practice, each ontology must therefore commit its vocabularies to certain reality [1]. Take the following RDF statement as an example (See section 2 for the syntactical convention):

```
_:x biopax-1:NAME 'Myoglobin'.
```

In the absence of the ontological commitment of *biopax-1:NAME*, the above statement can be used by a logic language to model any relations between $\_:x$ and "Myoglobin", among which the full name, short name, synonym, or primary identifier of an external database, etc., are all valid choices. But with *biopax-1:NAME* being committed to "the preferred full name" as defined in its URI, the above statement should allude only to $\_:x$'s naming relation to "Myoglobin".

Every ontological term, therefore, inherently carries two kinds of meanings (Fig. 1). The first kind is *extensional*. In logic and philosophy, an extensional definition formulates its meaning by specifying its extension, that is, every object that falls under the definition of the term. In the web, the extensional meaning of a URI is carried by the URI itself because the meaning is implied in the term's referred extensional entity. The extensional meaning of *biopax-1:NAME*, for instance, is simply the abstract concept of *"full name"*. The second kind of meaning of an ontological term is *intensional*. Intensional meaning is commonly expressed as a logic theory, i.e., a set of necessary and sufficient conditions, for the term to be satisfied. In the case of *biopax-1:NAME*, for example, its intensional meaning is the being of an *owl:DatatypeProperty* that can be used to associate a string-type data with one of the following entities: *entity, biosource, and*

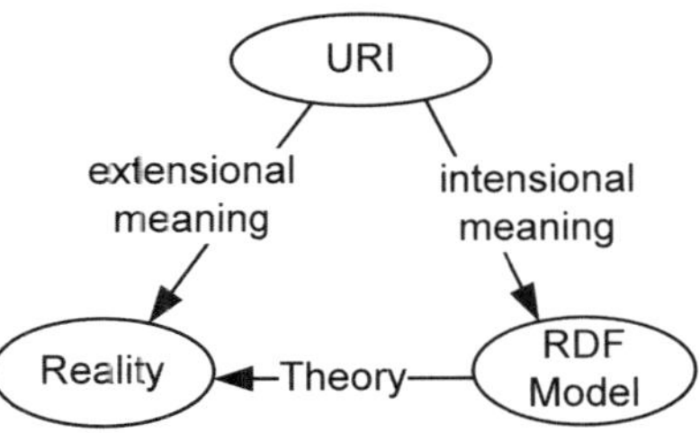

Fig. 1. The meaning of a URI

datasource. Using *biopax-1:NAME* to associate a *biopax:entity* with an integer value, for instance, would violate the intensional meaning of the *biopax-1:NAME*. Collectively, the extensional and intentional meanings of an ontological term build the term's conceptualization about an external reality [2].

Traditionally, an ontology is developed for the sharing of its intensional meanings; less care is taken to ensure the sharing of its extensional ones. This is understandable because, until the development of the web, the identity scope of ontological terms is typically restricted to individual files. When various conceptualizations are used across multiple ontologies, individual names must be either manually aligned or resort to external tool support like Ontolingua [3, 4]. The use of URI in the semantic web, however, should change this practice. Within an RDF document, the distinction between local and foreign identifiers becomes inconsequential. Hence, ontologies deployed in the web should be developed in a fashion that can ensure the maximal sharing of not only its intensional meanings but also its extensional ones. The importance of sharing the latter must not be taken lightly. Because rarely will an extensional entity need – let along it hardly can – be completely identified by a set of logic constraints, every extensional entity is likely to be explained by various logical theories. Sharing ontology's extensional meanings, *viz.* URIs, enables those consistent theories to be freely combined in an engineered system. In addition, it allows those inconsistent theories to still coexist and to be communicated at human level. Hence, when an ontology is treated as a shared engineered artifact [4], sharing URIs helps to lower the cost of system integration; when it is used as a social agreement [5], it helps to retain the ontological commitment of a term as different theories about the term continue to evolve.

The purpose of this article, therefore, is to introduce a few design principles and engineering techniques that can be used in practice to improve the sharing and reuse of ontological URIs. To illustrate the benefit of these techniques, we used the ontologies developed at BioPAX (http://www.biopax.org) as our use cases. The reason that BioPAX ontologies are chosen is two-fold. First, BioPAX ontologies exist at three different levels, providing us a retrospective frame on the problem. A detailed analysis of these three BioPAX ontologies will allow us to glimpse at the complexity of ontology development, which, in turn, will help us devise appropriate strategies to cope with them. Second, as the collaborative effort of numerous researchers from a number of institutes, the conceptualizations established in the BioPAX ontologies have become a valuable asset to the community. Hence, by analyzing the existing problems and finding solutions to avoid them in the future, we can further expand the applicability of this already valuable knowledge. A point that we want to stress here is that: it is easy – and understandably so – for people to take the provided problem analysis as a way to dismiss BioPAX ontologies. Thinking so, however, would misunderstand our intentions, which are to introduce ontology design principles and deployment techniques. We could easily build a set of trivial and imaginary ontologies to discuss the problem, but doing so would lessen the importance and urgency of these issues. Analyzing BioPAX, therefore, is aimed at offering constructive advices, as opposed to destructive criticisms, to the ongoing BioPAX development so that the resulting products could be more easily, efficiently, and broadly shared in the web.

2　Materials and Conventions

The three BioPAX ontologies used in this article are: Level-1 version 1.4, Level-2 version 1.0, and Level-3 version 0.9. Because of the different namespaces of the three levels of BioPAX ontologies, conceptualizations are compared in reference to the simple names of each URIs. In other words, if two BioPAX URIs have the same simple name, they will be assumed to denote the same extensional entity.

Qualified names as defined in [6] are used to shorten the URI notation. The URIs of all namespace prefixes used in this article except "biopax" are listed in Table 1. The prefix *biopax* is used to refer to a BioPAX concept in a general sense. It is worth noting that the namespace URI of the level-3 BioPAX ontology listed in Table 1 is not the official URI. Level-3 ontology was released informally through the BioPAX discussion mailing list. As the mailing list requires a registered account to access, we provided a copy at our own domain for the reader's convenience.

Table 1. Namespace Prefixes

Prefix	URI
biopax-1	http://www.biopax.org/release/biopax-level1.owl#
biopax-2	http://www.biopax.org/release/biopax-level2.owl#
biopax-3	http://dfdf.inesc-id.pt/ont/biopax-3#
owl	http://www.w3.org/2002/07/owl#
dc	http://purl.org/dc/elements/1.1/
o3	http://dfdf.inesc-id.pt/ont/o3#
[other]	http://dfdf.inesc-id.pt/ex/biopax/[other]#

All sample RDF statements will be written according to the syntax of Notation-3[7] with the namespace prefixes defined in Table 1. All RDF diagrams, such as Fig. 2, 3, and 5, are drawn with the graphical notation syntax of DLG^2 as defined in [8] as well as one of its extension defined in [9].

3   Analysis of BioPAX Ontology

BioPAX ontologies gradually evolved over time. Level-1 ontology defined the usage of 76 terms, with additional 34 and 54 introduced at level-2 and 3, respectively. Each of the three ontologies, however, is governed by a unique namespace, making the data instances described at one level of ontology unable to interoperate with those at another. To make the problem easily understandable and presentable, let's take *biopax:NAME* as an example. Consider the following two statements.

```
_:x biopax-1:NAME 'Myoglobin'.
_:x biopax-2:NAME 'Myoglobin'.
```

From a machine's standpoint, the above two statements, in fact, entail two entirely different theories about the resource "_:x". To promote interoperability, a newly developed ontology, such as BioPAX level-2/3, should in principle reuse as much the conceptualizations defined in the existing ones, such as BioPAX level-1, as possible. Obviously, the problem can always be solved with a post-integration approach. For instance, the two versions of biopax:NAME can be easily aligned using the following assertion.

```
biopax-1:NAME owl:sameAs biopax-2:NAME.
```

This integration approach, however, suffers from two drawbacks. First, it is potentially expensive because, for every k URIs of the same conceptualization, $k(k-1)/2$ statements must be made. As careful ontology engineering can easily avoid the problem, the integration approach should be taken as the last resort to integrate legacy semantic objects, such as those deployed in relational databases[10]. Second, the integration approach may not be possible when two conceptualizations logically contradict each other (see section 3.1.2).

The seemingly trivial problem encountered by the *NAME* conceptualization, in fact, elucidates a fundamental issue with regard to the knowledge engineering in the web. That is: how can the URI of an ontological term be maximally shared within the web? As ontological URIs denote conceptualizations, which can only be shared if they are logically *consistent* with each other, we should first investigate ontology's compatibility issues before evaluating the sharing capability of BioPAX ontologies. Strictly speaking, "consistency" should be used in place of "compatibility" to describe ontology's sharing capability. However, as OWL has defined two properties – *owl:incompatibleWith* and *owl:backwardCompatibleWith* – to describe ontologies' inter-relationship, it is in the best interest of the web community as a whole for us to align our terminologies with OWL.

3.1 Ontology Compatibility

Traditionally in computer science, product compatibility is defined with regard to their functional interface. A new version of software, for instance, is considered backward compatible if it can take the place of an older one in terms of fulfilling the existing functionalities. Ontology development is, however, different: it concerns data rather than functionalities; and it aims at sharing and reuse rather than replace. A new ontology is backward compatible if all data instances described in the new ontology are also valid data instances of existing ones.

Between ontology's two kinds of meanings, the extensional meanings should not incur much, if any, ontology's incompatibility. Extensional meaning is consumed by humans and their compatibilities are maintained through social agreement, which should in principle be very stable. The cause of ontology's incompatibility, therefore, should mainly come from the specification of its intensional meanings. As the intensional meanings of ontological terms are typically defined in logic languages, one ontological term is compatible with the other only if they are logically consistent. In the subsequent sections, we used a few examples from BioPAX ontologies to discuss the issue.

3.1.1 Compatible Use Case

Consider the definitions of *conversion* class in BioPAX ontologies. At level-1, *conversion* subclasses directly from *interaction* (Fig. 2a). But, at level-2, it does so indirectly from *physiccalInteraction* – a class that was newly introduced into BioPAX at level-2 (See Fig. 2b). Because all instances of *biopax-2:conversion* would be valid instances of *biopax-1:conversion*, the level-2's definition of *conversion* is compatible with that of the level-1.

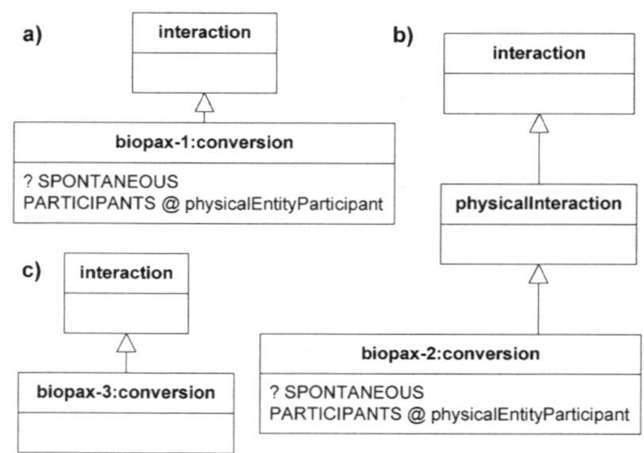

Fig. 2. BioPAX's conversion class. (a) Level-1 definition (b) Level-2 definition (c) Level-3 definition.

The converse, however, is not true because a *biopax-1:conversion* is not necessarily a *physical-Interaction*, and therefore, a *biopax-2:conversion*. By the same reasoning, the *conversion* defined at level-3 (Fig. 2c) is not compatible with either *biopax-1:conversion* or *biopax-2:conversion*. The removal of the property constrains on SPONTANEOUS and PARTICIPANTS makes *biopax-3:conversion* a more subsuming class than its counterparts in level-1/2. For

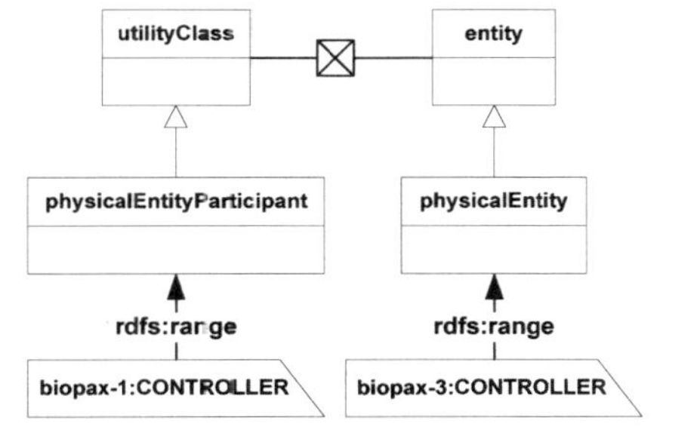

Fig. 3. CONTROLLER definitions in BioPAX

example, a *conversion* instance with two SPONTANEOUS properties would be a valid *conversion* at level-3 but an invalid one at level-1/2.

Strictly speaking, an ontology term is either compatible with another one or it is not. But as the purpose of this article is to gain an insight into ontology development through the analysis of BioPAX ontologies, we designated the case similar to *biopax-3:conversion* as a *reverse compatible* case, in the sense that the compatibility is achieved in a reversed direction with reference to the progression of ontology development. The word "incompatibility", therefore, is reserved for those ontological terms whose conceptualizations are not compatible in either direction.

3.1.2 Incompatible Use Case

The conceptualization of *CONTROLLER* property is a typical incompatible use case. At all three levels of BioPAX ontologies, *CONTROLLER*'s domain is defined to be *control*, but its range was changed at level-3 from *physicalEntityParticipant* to *physicalEntity* (Fig. 3). Because the latter two classes are respectively extended from two disjoint classes – *utilityClass* and *entity*, *biopax-3:CONTROLLER* is incompatible with either *biopax-1* or *biopax-2:CONTROLLER*.

It is worth noting that the concept of *physicalEntityParticipant* is left undefined at level-3. Owing to the informal status of the level-3 ontology that was used in this analysis, we are unsure whether the change is made by intention or by accident. In either case, nevertheless, the rationale behind the use case is valid because, in science, incompatible or contradicting theories about the same reality often coexist or emerge over time.

A milder case of incompatibility can be seen from the definition of *FEATURE-TYPE* property. At level-2, the property's domain is set to be *sequenceFeature*, but at level-3 it is changed to be *entityFeature*. Both *sequenceFeature* and *entityFeature* subclass from *utilityClass*, but their inter-relationship has yet to be defined in BioPAX. Logically, the two versions of *FEATURE-TYPE* are not inconsistent, but conceptually they reflect two different modeling approaches to the same entity. Sharing biopax-2:*FEATURE-TYPE* with biopax-3: *FEATURE-TYPE* will not improve data's interoperability because the same set of data instances would be interpreted by two completely different set of modeling primitives. In this article, this kind of definition was treated as an incompatible case because neither definition subsumes the other.

For the same reason, changing the property type from *owl:DatatypeProperty* to *owl:ObjectProperty* or *vice versa* is also considered an incompatible change, albeit its only consequence is the increase of inference complexity. A case in point is *biopax:KEQ*, which is an *owl:DatatypeProperty* at level-1 but an *owl:ObjectProperty* at level-2/3.

3.2   Compatibility of BioPAX Ontologies

Using the above defined criteria, we analyzed the inter-compatibility of ontological terms among three levels of BioPAX ontologies. The result is shown in Fig. 4. A detailed term-by-term analysis is provided at [11].

To simplify the analyzing process, two contributing factors – disjoint statement and term dependency – were disregarded because a full account of these factors would significantly increase the workload of the analysis without necessarily changing the delivered message.

The disjoint statement was disregarded due to its abundant use in BioPAX ontologies. A simple count of level-3 ontology, for example, revealed a total of 357 *owl:disjointWith* statements. Such abundance, together with the varying class hierarchy at different levels of BioPAX ontology, made the analysis of owl:dis-joint statements quite a time consuming process. To reduce the workload, all disjoint statements were therefore excluded from the evaluation. Nevertheless, changing *owl:dis-jointWith* statement will have an effect on a term's compatibility. Adding a disjoint statement will make a class more logically restrictive, and removing one will make it more logically general, than its original form. A case in point is the *biopax-3:pathway*, which has an extra disjointed class – *referenceEntity* – compared with its counterparts at level-1/2.

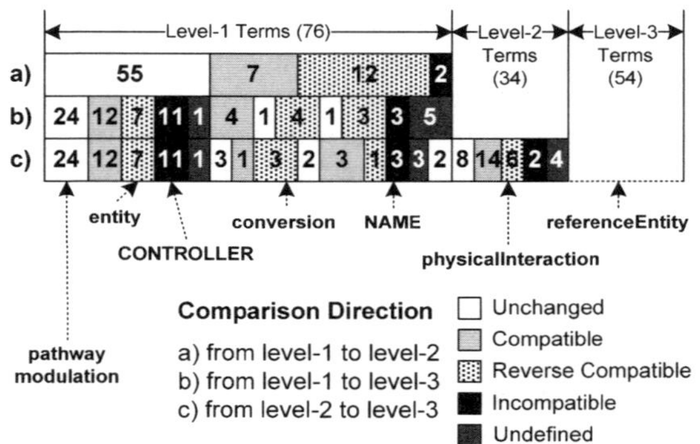

Fig. 4. Compatibility Analysis of BioPAX Ontologies. The compatibilities of the same term among three levels of ontologies are vertically aligned. The few example cases discussed in the text are illustrated. The numerical value inside each box shows the number of BioPAX terms with the similar compatibilities, which type is indicated by the boxes' filled pattern.

The term's inter-dependency was disregarded owing to the difficulties of evaluation. Take the *modulation* class as an example. At all three levels of BioPAX ontologies, *modulation* is defined to be a subclass of *control* with no more than one *CONTROLLER*, and with all CONTROLLED property coming from *catalysis*. But as the CONTROLLER's definition at level-3 logically contradicts those at level-1/2 (see last section), the *modulation's* dependency on the CONTROLLER property would make their definitions contradicting as well. In a logic system, a single contradiction would invalidate the entire theory. Even if we do not take this position during the evaluation, the interdependency of ontological terms will eventually lead us to the same conclusion. As seen in Fig. 3, the CONTROLLER's definition is ultimately related to *entity* and *utilityClass* – the very two top classes of BioPAX ontologies. Counting dependencies of BioPAX term will make *entity* and *utilityClass*, and therefore the entire ontology, a contradiction. In short, taking the full account of term dependency would prevent us from making a more detailed and meaningful analysis. In the provided analysis, the compatibility was therefore evaluated solely based on their syntactic definitions; all dependency incurred changes were disregarded.

4   Methods to Improve Ontology's Sharing

The current practice of BioPAX ontology development is not optimal in terms of sharing and reusing existing conceptualizations. First, at each level of BioPAX ontologies, the same concept is renamed under a different namespace, making the sharing of their extensional meaning difficult. Second, the ontologies are presented in a monolithic fashion, making the sharing of their intensional meaning difficult as well.

Ideally, an ontology should be developed in an incremental fashion. Useful and relevant conceptualizations developed previously should be imported into the new ontology, where they can be logically refined with newly introduced conceptualizations. Take the conceptualization of *pathway* and *modulation* as an example. Their definitions (sans disjoint statements) remain unchanged throughout all three levels of BioPAX ontologies. In principle, both *pathway* and *modulation* need only to be defined once, e.g., at level-1, and then reused, e.g., at level-2/3, *via* a simple *owl:imports* statement. The same importing approach can also be applied to the compatible terms, such as *conversion*, whose level-1 definition can be imported into level-2, where additional subclass statements can be made to realign the class hierarchy in reference to the newly introduced class – *physicalInteraction*. Such an incremental approach allows both the URI and the logic definitions of a conceptualization to be shared. Interoperability will be improved as data instances developed against the same ontology can be unambiguously understood on the same ground without additional engineering effort. Furthermore, it would also reduce the cost of ontology development and maintenance because the same statements no longer need to be redundantly defined.

However, if ontologies are developed in a monolithic fashion, the incremental approach is unlikely to take place. Consider the development of BioPAX ontologies. Conceptualizations that would undergo different compatibility changes are physically bound in a single document (Fig. 4). The conceptualization of *modulation, pathway,* and *conversion* etc., for instance, were bundled up with that of *NAME* and

CONTROLLER. Because, unlike *conversion*, which underwent a compatible change at level-2, the *biopax:NAME* made a reverse compatible change at level-2, suggesting that, if the conceptualization is to be shared, the import must be made from level-2 to level-1, but not *vice versa*. The *CONTROLLER*, on the other hand, underwent an incompatible change at level-3, suggesting that neither definition can be imported into the other.

Since RDF does not yet support ontology modulization techniques, such as named graph [12] or C-OWL [13], statements in an RDF document cannot be selectively imported into another. To improve the sharing and reuse of existing conceptualizations, ontologies must, therefore, be carefully designed and deployed. First, large, monolithic ontologies should be broken up into sets of small and modular ones. Second, ontology's inter-dependency must be carefully structured in a manner that it avoids tight coupling. In the subsequent sections, we will introduce a few design principles and engineering techniques that may help in this regard.

4.1 Ontology Design Principles

4.1.1 Minimal Ontological Commitment

The principle of minimal ontological commitment (PMOC) was proposed by Gruber [14]. Here we quote,

An ontology should make as few claims as possible about the world being modeled, allowing the parties committed to the ontology freedom to specialize and instantiate the ontology as needed. Since ontological commitment is based on consistent use of vocabulary, ontological commitment can be minimized by specifying the weakest theory (allowing the most models) and defining only those terms that are essential to the communication of knowledge consistent with that theory.

Gruber proposed PMOC along with five other sound ontology design principles in the same article. Only PMOC is chosen here because it is the most pertinent to the ontology's sharing and reuse. Principles of similar essence have in fact been echoed in other research areas. For instance, *the principle of least effort* has been used to theorize the user behavior during information search, and *the rule of least power* has been proposed by W3C as the guidance for language design [15].

A case in point is BioPAX's definition of *NAME*. At level-1, the domain of *NAME* is defined to be the union of *entity*, *bioSource*, and *dataSource*. At level-2, a new class – *sequenceFeature* – is added to the domain, making *biopax-2:NAME* a reverse compatible definition of *biopax-1:NAME*. At level-3, however, the *NAME*'s domain is redefined to be the union of *bioSource*, *pathway*, *physicalEntity*, *and referenceEntity*, making it incompatible with either level-1 or 2's definition. However, as things that can have a *full name* – the NAME's extensional meaning – are not restricted to those entities defined in BioPAX, the conceptualizations of *NAME* have apparently over-committed its intended use. To follow the advice of PMOC, the domain of *NAME* should not be constrained at all. At most, if so desired, it can be constrained to the union of *entity* and *utilityClass* – the two top classes of BioPAX ontologies. Such a design would stabilize the *NAME*'s conceptualization, which, in turn, would allow the definition be easily shared by all three levels of BioPAX ontologies.

4.1.2 Granularity Separation

The PMOC, however, should not be taken in a rigid and extreme sense because, otherwise, there should be only top ontologies like Suggested Upper Merged Ontology (SUMO) [16] but nothing else. Most ontologies are developed to carry a specific application task, which should be used as the context for PMOC to be meaningfully applied. Take the conceptualizations of *conversion* as an example (See Fig. 2). Although only the level-3's definition satisfies PMOC, it is hard to fault the design of level-1 and 2's because the coarse semantic granularity of level-3's definition may not suit the need of the targeted application.

There is in general a tradeoff between an ontology's expressiveness and its shareability. A coarse grained ontology carrying only a few logical constraints is easier to be shared but less powerful to constrain an application's behavior. A fine grained ontology bearing a rich set of axioms, on the other hand, is more logically clear in directing specific application behaviors but less convenient and more expensive to be integrated.

Obviously, there should be all kinds of ontologies conceptualized at every grain of granularity. But the point that we want to stress here is: ontologies of different semantic granularities should be separately developed and deployed. This is, as we shall name it, *the principle of granularity separation (PGS)*. For instance, had the development of level-1 BioPAX followed the PGS, the general form of *conversion* definition as that of level-3 would have been separately developed at level-1, preventing it from confounding the sharing of other conceptualizations at level-3.

The PGS further implies that a fine grained ontology should be developed within the framework of a coarse grained top ontology. Starting from a top ontology is, in fact, almost always a good strategy [2, 17]. A top ontology usually carries less logical constraints; its conceptualizations are more general and stable, which make them easier to be shared as consensus among large communities of users. Furthermore, a top ontology usually contains the general structural information, which can help users to systematically identify and partition the targeted knowledge domain, against which detailed application ontologies can be developed and integrated.

4.1.3 Orthogonal Domain

Though not explicitly defined in the ontology field, orthogonal separation has been a well received concept in computer science. In both software [18] and database design [19], orthogonality principle has been used as a key guideline to improve system's expressive power and reusability. The very first architecture principle of the Web is, in fact, *the principle of orthogonal specifications* because it helps to increase the flexibility and robustness of the Web [20]. Here we propose *the principle of orthogonal domain (POD)* with regard to ontology development.

Semantically orthogonal conceptualizations should be defined in separate ontologies.

Consider, for example, the relationship between *entity* and *name*. A biological pathway entity is a *biopax:entity* regardless if it has a name or not. Conversely, whether an entity has a *name* or not bears no relations to its nature as a *biopax:entity* or not. Of course, developing the conceptualization of *NAME* in BioPAX can hardly be blamed because to label a *biopax:entity* does require a naming concept. But, deploying the conceptualizations of *NAME* and *entity* in the same ontology should be criticized because it hinders, if not prohibits, alternative conceptualizations from being applied. For instance, the conceptualization of *NAME* is, in fact, identical to that of *dc:title* defined in the Dublin Core Metadata Initiative (DCMI) [21]. As the latter is a more thoroughly developed and shared community standard, *biopax:NAME* should eventually be replaced by *dc:title* to improve the overall data interoperability of BioPAX data. The physical binding of *entity* and *NAME* in the same document makes the replacement difficult to take place.

To follow the POD, the conceptualizations of *entity* and *NAME* should be separately specified in two different ontologies. There are two benefits of such a separation. First, separation reduces the sharing cost of each term, which in turn increases their shareabilities. In its current deployment form, for instance, *biopax:NAME* is unlikely to be used by someone, say a physicist, who has no interest in biology – not just because the *NAME*'s domain is so restricted, but more so because importing the *NAME*'s conceptualization would also import a few hundreds of assertions that are completely irrelevant to the physicist's tasks. Between increasing data interoperability and reducing computation cost, the physicist may be forced to choose the latter because a few importing of concepts similarly deployed as *biopax:NAME* would easily render his application into an unmanageable state. But if the conceptualization of *NAME* is separately deployed, the physicist would be more willingly to use it because doing so not only improves his data's interoperability but also reduces his development and maintenance cost.

Second, domain separation allows different conceptualizations to evolve independently without interfering with each other. In the field of knowledge engineering, there is an often encountered problem – the interaction problem, where "representing knowledge for the purpose of solving some problem is strongly affected by the nature of the problem and the inference strategy to be applied to the problem [22]". During the design of Chemical Markup Language (CML) [23], for instance, its designers have found "that many components with a 'chemical content' did not require chemical concepts for their implementation" [24]. Since one cannot, especially at the beginning of the Semantic Web, expect the presence of all kinds of ontologies available to his use, one must develop ontologies in both his familiar and his unfamiliar areas. Domain separation allows domain experts to divide their labors according to their area of expertise. A full scaled ontology can be developed in their familiar domains and makeshift solutions can be used to handle the areas that are out of their elements. More importantly, all these can be done without worrying that the latter's informality and less popularity may prevent the sharing and reuse of the former. In the next section, we introduce how such separation should be engineered in the semantic web.

4.2 Ontology Engineering

As conceptualizations are encoded and delivered in engineer artifacts, conceptual separations can only be realized if ontologies are engineered to be physically independent from each other. Consider the deployment of *entity* and *NAME* shown in Fig. 5a. At first glance, the two conceptuali-zations appeared to be separated because they are specified under two different namespaces "http://dfdf.inesc-id.pt/ex/biopax/a" and "http://dfdf.inesc-id.pt/ex/biopax/n", re-spec-tively. But a careful look will reveal that the separation is only half complete because *a:entity* is still tightly coupled with *n:NAME*, albeit not *vice versa*. A complete separation of the two conceptualizations should be deployed as shown in

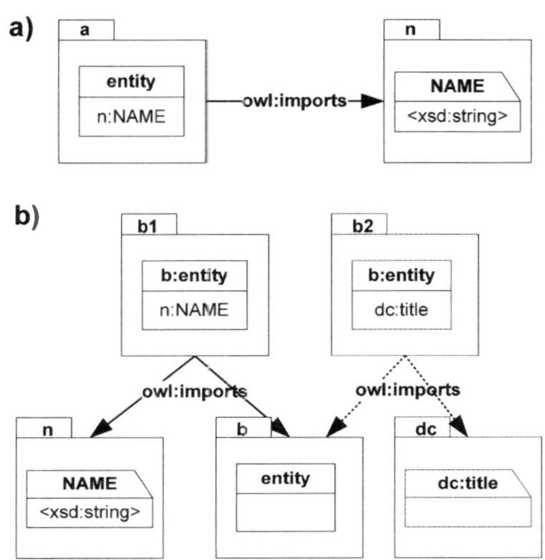

Fig. 5. Ontology Deployment Strategy. (a) and (b) shows two different deployment strategies. The dotted shapes show the potential changes in the future.

Fig. 5b, where *entity* and *NAME* are respectively specified at ontology "http://dfdf.inesc-id.pt/ex/biopax/b" and "http://dfdf.inesc-id.pt/ex/biopax/n". The two in-dependent conceptualizations are collectively used at ontology "http://dfdf.inesc-id.pt/ex/biopax/b1".

The advantage of a complete separation is that concepts defined in one ontology can be used independently of those in the other. This independence firstly reduces the sharing cost and, therefore, increases shareability. Secondly, it allows existing conceptualization to gracefully evolve over time without tampering the compatibility of the others. For instance, the *entity*'s conceptualizations in both "http://dfdf.inesc-id.pt/ex/biopax/b1" and "http://dfdf.inesc-id.pt/ex/biopax/b2" evolve from *b:entity*. But in ontology *b2*, *dc:title* is used in place of the *n:NAME* in '*b1*'. Owing to the complete independence between ontologies *b* and *n*, *b2* can coexist with *b1* so that the migration from one to another can be carried out gracefully. If, on the other hand, the *entity* is deployed as shown in Fig. 5a, the independent evolution of *entity* becomes impossible.

4.2.1 Ontology Classification

To facilitate the discussion on ontology's engineering issue, we defined an Ontology of ontologies (O3) [25] that classifies ontologies along few semantic dimensions. First, in O3, we made a distinction between ontologies that define a logic language, such as RDFS and OWL etc., from the rest, such as BioPAX and DCMI [21]. The former is defined as instance of *o3:Logic* and the latter of *o3:DomainOntology*. Second, we classified ontologies by their physical dependencies. If an ontology's conceptualizations are independent of the others, i.e., if the ontology does not import another *o3:DomainOntology*, it is defined to be an *o3:Local*. Otherwise, it is an *o3:Complex*. Third, we categorize ontologies according to the kinds of meanings that they carry. An ontology carrying only the extensional meanings of its terms is an *o3:Vocabulary*; one carrying only the intensional meanings is an *o3:Theory*; otherwise, it is an *o3:ConcreteOntology*. Different combinations of these basic ontology classes can lead to the definitions of *o3:LocalOntology*, *o3:ComplexOntology*, etc. An important class worthy of special mention is *o3:Profile*. An *o3:Profile* is an *o3:ComplexTheory* with characteristic RDF statements: all *o3:Profile*'s statements are made of concepts initially deployed elsewhere, meaning that the sole functionality of an *o3:Profile* is to make joint use of independent conceptualizations. Fig. 6 is a Venn diagram of the relationships among several ontology classes relevant to this discussion. More detailed information of those classes can be found at [25].

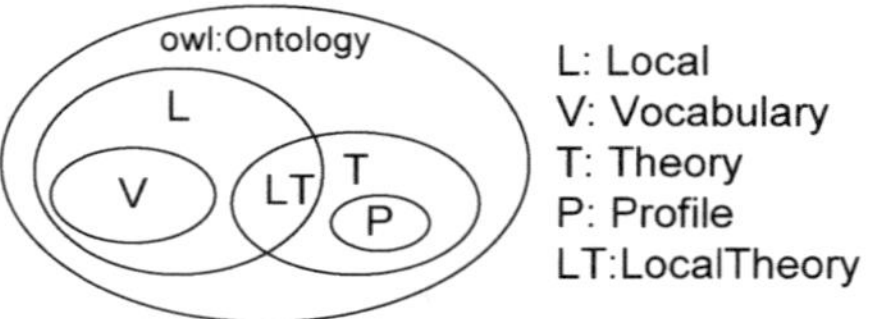

Fig. 6. Ontology classification

4.2.2 Normalized Ontological System

With the conceptualizations established in O3, we can now define two ideal forms of ontology deployment.

- *Intensionally normalized form (INF)*: An ontological system is in intensionally normalized form if it consists of only *o3:Local* and *o3:Profile*.
- *Extensionally normalized form (ENF)*: An ontological system is in extensionally normalized form if it consists of only *o3:Vocabulary* and *o3:Theory*.

INF and ENF are not mutually exclusive. They can be conjunctively applied to an ontological system, resulting in a system that consists of only *o3:Vocabulary*, *o3:LocalTheory*, and *o3:Profile*.

By the above definition, the system deployed in Fig. 5a is an anomaly with regard to INF because ontology *a* is an *o3:Complex*. On the hand, the system deployed in Fig. 5b is in INF because ontologies *b, n* are *o3:Local* and *b1, b2* are *o3:Profile*. As we have discussed earlier, the advantage of an INF system is that ontologies' inter-dependency is carefully managed so that each ontology has the most shareability.

But deploying ontologies in INF is still insufficient to maintain the URI's stability under certain circumstances. A case in point is the conceptualization of *biopax:CONTROLLER* (See Fig. 3), whose two conceptualizations, if shared, leads to a logical contradiction. Thus, if the logical constraints of *CONTROLLER* are bound with the instantiation of its URI, two different URIs must be used to denote the two kinds of *CONTROLLER*. However, the two types of CONTROLLER differ only in their intensional meanings; their extensional meanings are still the same, implying that their URIs do not have to be different. To maintain the URI's stability, CONTROLLER can be first developed in an *o3:Vocabulary*, where its extensional meanings are first to be specified. This vocabulary URI can then be subsequently used in various *o3:Theory* to define its intensional meanings. In the history of science, a fundamental change in our conceptualization about an external entity was rarely, if ever, accompanied by a change in the words that are used to refer to the entity. For instance, Copernican revolution changed our conceptualization about the earth, but not the usage of the word "earth". The same principle, we think, should also be applied to ontological systems as well, and deploying ontologies in ENF will help in this regard. An *o3:Vocabulary* only establishes the identities of resources in the web but does not make any logical assertions about them. URIs defined in an *o3:Vocabulary*, therefore, can be shared in any logic theories, regardless if they contradict each other or not.

5 Conclusion

The web is an open, decentralized system, in which communication is carried out through sharing resources. Ontologies are no exception. They are web resources and they are to be shared. Good ontologies thus should be developed to have the maximal shareability. Of course, correct conceptualization requires an ontology's shareability. But other factors would also contribute. The first one is the size of an ontology. An ideal ontological system should not be comprised of a single, consistent, and comprehensive ontology. Putting aside the debate on whether such an ontology is attainable, we think that, even if it were, it may not be desirable. The size of the knowledge base is always an important performance factor for an ontology driven application [26]. Using a few concepts from a galaxy-like ontology will not be a sensible approach under most circumstances. Most domain applications would have a well defined local task that rarely demands concepts beyond their domain knowledge. For instance, a BioPAX-driven program is unlikely to be interested in whether a *pathway* or a *conversion* is aligned to the "Physical" or "Abstract" as defined in SUMO. Forcing their alignment into a single ontology can only tax the program with unnecessary computation cost. This is, however, not to say that SUMO, or any other top ontology of the similar nature, is not useful. On the contrary, we believe that the opposite is true. A well defined top ontology always serves a valuable conceptual framework for guiding the design of domain ontologies. But, conceptually alignment does not have to be always realized in physical implementation. The broad spectrum of application needs demands that all kinds of ontologies be developed and be at every grain of semantic granularity. Moreover, each of them should exist independently, while readily to be assembled into a coherent system.

The second contributing factor to an ontology's shareability is its stability. As no useful engineered system can be developed to varying specifications and no agreement can be made to varying subjects, an ontology, once developed and deployed, must seldom, if ever, change. But things always change with time; ontologies are no exception. On one hand, scientific progress can change our conceptualizations about an extensional entity. On the other hand, technological advancement and social interaction can change the way that a particular problem is solved. The challenge, therefore, lies in how to find the balance between a system's stability and its adaptability. As shown in this article, the monolithic approach taken by the BioPAX ontologies cannot meet the challenge. In a monolithic system, change must take place in an all-or-none fashion. Because all conceptualizations are physically bound together, a partial change must be made in the original ontology or a completely newly developed one. Both approaches have serious drawbacks. The former risks the danger of breaking an existing application that depends on the previous conceptualizations; the latter impedes data interoperability by using different set of URIs for the same conceptualization.

"A new scientific truth", as Max Planck has judiciously stated in [27], "does not triumph by convincing its opponents and making them see the light, but rather because its opponents eventually die, and a new generation grows up that is familiar with it." Although an ontological system engineers – as opposed to discover – scientific truth, its use and sharing in the semantic web should nevertheless follow the same principle. A new ontology should not be put into use by replacing an older one. Rather, it should reuse or compete with the older one if reuse is not possible. The ultimate winner is chosen by the users rather than by the designers of an ontology. In this sense, the best ontologies can be defined to be the ontologies that are mostly shared, and the worst are those that are seldom linked. There is perhaps no such thing as the ideal ontology but only the ideal ontological system that is capable of fostering an ideal one. We hope that the few design principles and engineer techniques introduced in this article may offer help toward building such a system.

Acknowledgement. This work was developed while the first author was supported by a Portuguese FCT (Fundação para a Ciência e a Tecnologia, Portugal) Post-doctoral Fellowship grant with reference SFRH/BPD/34997/2007. The work is also partially supported by FCT supported PTDC program (project DynaMo - PTDC/EEA-ACR/69530/2006) and by the NHLBI Proteomics Initiative through contract N01-HV-28181 to the Medical University of South Carolina, PI. D. Knapp, specifically as refers to the work of its bioinformatics core (core C) and mathematically modeling project (project 7), as well as by its administrative center, separately funded by the same initiative to the same institution, PI. M.P. Schachte.

References

1. Guarino, N., Carrara, M., Giaretta, P.: Formalizing Ontological Commitments. In: National Conference on Artificial Intelligence (AAAI 1994). Morgan Kaufmann, Seattle (1994)
2. Guarino, N.: Formal Ontology and Information Systems. In: Formal Ontology in Information Systems. IOS Press, Amsterdam (1998)
3. Farquhar, A., Fikes, R., Rice, J.: The Ontolingua Server: A Tool for Collaborative Ontology Construction. KSL-96-26. Knowledge Systems Laboratory, Department of Computer Science, Stanford University, Stanford, CA (1996),
   `http://ksi.cpsc.ucalgary.ca/KAW/KAW96/farquhar/farquhar.html`
4. Gruber, T.: A translation approach to portable ontology specifications. Knowledge Acquisition 5, 199–220 (1993)
5. Gruber, T.: Interview Tom Gruber. SIGSEMIS Bulletin 1, 4–9 (2004)
6. Bray, T., Hollander, D., layman, A., Tobin, R.: Namespaces in XML 1.0, 2nd edn: W3C Recommendation (2006),
   `http://www.w3.org/TR/REC-xml-names/#ns-qualnames`
7. Berners-Lee, T.: Notation 3 (1998), http://www.w3.org/DesignIssues/Notation3.html

8. Wang, X., Almeida, J.S.: DLG2- A Graphical Presentation Language for RDF and OWL (2005), `http://www.charlestoncore.org/dlg2/`
9. Wang, X., Oliveira, A.L.: DLG2 Extensions for Property Constrains (2007), http://dfdf.inesc-id.pt/dlg2
10. Kashyap, V., Sheth, A.: Semantic and Schematic Similarties between Database Objects: A Context based Approach. VLDB Journal 5, 276–304 (1996)
11. Wang, X., Almeida, J.S., Oliveira, A.L.: Compatability Analysis of Terms in BioPAX Ontologies (2007), `http://dfdf.inesc-id.pt/tr/biopax-analysis`
12. Carroll, J.J., Bizer, C., Hayes, P., Stickler, P.: Named graphs, provenance and trust. In: Proceedings of the 14th international conference on World Wide Web. ACM Press, Chiba (2005)
13. Bouquet, P.G.F., van Harmelen, F., Serafini, L., Stuckenschmidt, H.: Contextualizing Ontologies. Journal of Web Semantics 1 (2004)
14. Gruber, T.R.: Toward Principles for the Design of Ontologies Used for Knowledge Sharing. Knowledge Systems Laboratory. Stanford University (1993)
15. Berners-Lee, T., Mendelsohn, N.: The Rule of Least Power (2006), `http://www.w3.org/2001/tag/doc/leastPower.html`
16. Niles, I., Pease, A.: Towards a Standard Upper Ontology. In: Welty, C., Smith, B. (eds.) Proceedings of the 2nd International Conference on Formal Ontology in Information Systems (FOIS 2001), Ogunquit, Maine (2001)
17. Rector, A.: Modularisation of Domain Ontologies Implemented in Description Logics and related formalisms including OWL. In: Knowledge Capture, pp. 121–128. ACM, Sanibel Island, FL (2003)
18. Atkinson, M., Morrison, R.: Orthogonally persistent object systems. The VLDB Journal 4, 319–402 (1995)
19. Date, C.J., McGoveran, D.: A New Database Design Principle. Database Programming & Design 7, 46–53 (1994)
20. W3C Technical Architecture Group: Architecture of the World Wide Web, Volume One. In: Jacobs, I., Walsh, N. (2004), `http://www.w3.org/TR/webarch/`
21. DCMI: DCMI Metadata Terms, http://dublincore.org/documents/dcmi-terms/
22. Bylander, T., Chandrasekaran, B.: Generic tasks in knowledgebased reasoning: The right level of abstraction for knowledge acquisition. In: Gaines, B.R., Boose, J.H. (eds.) Knowledge Acquisition for Knowledge Based Systems. Academic Press, London (1988)
23. Murray-Rust, P., Rzepa, H.S.: Chemical markup, XML and the World-Wide Web. 2. Information objects and the CMLDOM. J. Chem. Inf. Comput. Sci. 41, 1113–1123 (2001)
24. Murray-Rust, P., Rzepa, H.S.: STMML. A markup language for scientific, technical and medical publishing. Data Science Journal 1, 1–65 (2002)
25. Wang, X., Almeida, J.S., Oliveira, A.L.: Ontology of Ontologies (2008), `http://dfdf.inesc-id.pt/ont/o3`
26. Heinsoh, J., Kudenko, D., Nebel, B., Profitlich, H.-J.: An Empirical Analysis of Terminological Representation System. Artificial Intelligence 68, 367–397 (1994)
27. Planck, M.: Scientific Autobiography and Other Papers. Philosophical Library, New York (1949)

Exploiting Ontology Structure and Patterns of Annotation to Mine Significant Associations between Pairs of Controlled Vocabulary Terms

Woei-Jyh Lee[1], Louiqa Raschid[1], Hassan Sayyadi[1], and Padmini Srinivasan[2]

[1] University of Maryland, College Park, MD 20742, USA
{adamlee,louiqa}@umiacs.umd.edu, sayyadi@cs.umd.edu
[2] The University of Iowa, Iowa City, IA 52242, USA
padmini-srinivasan@uiowa.edu

Abstract. There is significant knowledge captured through annotations on the life sciences Web. In past research, we developed a methodology of *support* and *confidence* metrics from association rule mining, to mine the *association bridge* (of *termlinks*) between pairs of controlled vocabulary (CV) terms across two ontologies. Our (naive) approach did not exploit the following: implicit knowledge captured via the hierarchical *is-a* structure of ontologies, and patterns of annotation in datasets that may impact the distribution of parent/child or sibling CV terms. In this research, we consider this knowledge. We aggregate *termlinks* over the siblings of a parent CV term and use them as additional evidence to boost *support* and *confidence* scores in the associations of the parent CV term. A weight factor (α) reflects the contribution from the child CV terms; its value can be varied to reflect a variance of confidence values among the sibling CV terms of some parent CV term. We illustrate the benefits of exploiting this knowledge through experimental evaluation.

Keywords: annotation, controlled vocabulary (CV) terms, generalized association rule mining, support and confidence, life sciences link (*LSLink*).

1   Introduction

The biomedical enterprise has generated an abundance of data that is captured using annotated and hyperlinked records in the life sciences Web. Records in each resource are typically annotated with terms from controlled vocabularies (CVs) or ontologies, forming a rich Web of knowledge. Consider a simplified Web of three publicly accessible resources Entrez Gene [1], OMIM [2] and PubMed [3], in Figure 1. Data records in each resource are annotated with terms from multiple CVs. The hyperlinks between data records in any two resources form a relationship between the two resources, represented by a (virtual) link. Thus, a record in OMIM, annotated with SNOMED terms [4] has multiple links to gene records in Entrez Gene, annotated with GO terms [5]; gene records further have hyperlinks to multiple records in PubMed annotated with MeSH terms [6].

A. Bairoch, S. Cohen-Boulakia, and C. Froidevaux (Eds.): DILS 2008, LNBI 5109, pp. 44–60, 2008.
© Springer-Verlag Berlin Heidelberg 2008

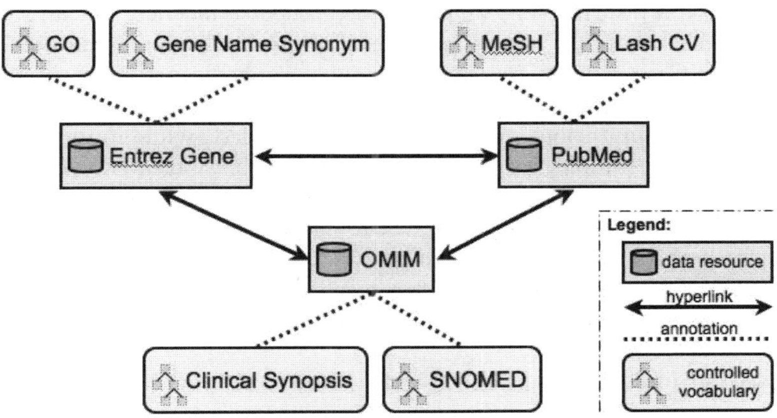

Fig. 1. Web of Entrez Gene, OMIM and PubMed Resources

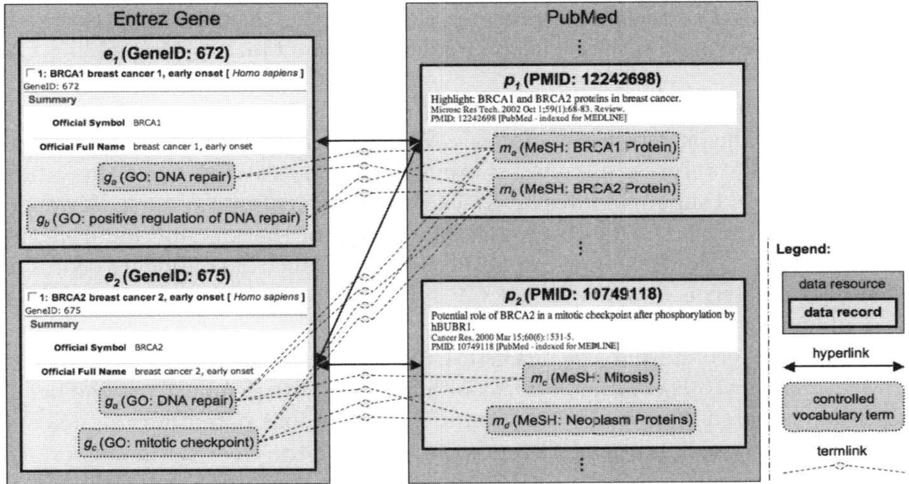

Fig. 2. Example hyperlinks between Entrez Gene and PubMed

A background *LSLink* (Life Sciences Link) dataset composed of *termlinks* (to be defined in Section 2.1) is generated after executing a protocol to follow hyperlinks and to extract annotations; details are provided in [7]. Example hyperlinked records are shown in Figure 2. Each *termlink* associates a pair of CV terms, and contributes to an *association bridge* (reflecting a connection of two CV terms) across two CVs or ontologies. A user dataset is a subset of the background dataset. In prior research, we mined the *association bridge* of *termlinks* of user datasets, to discover potentially new knowledge that is both meaningful and not well known a priori. Using *support* and *confidence* metrics, we can rank the pairs of associations of CV terms and identify potentially significant pairs. User validation confirmed that a majority of highly ranked pairs were meaningful. Several pairs were unknown and might lead to actionable knowledge [7].

There are two limitations of our prior research. First, while mining the association bridge of termlinks between pairs of CV terms, we treated each CV term (of the CV or ontology) independently. For example, *is-a* is a key relationship that exists amongst terms of a single vocabulary. Intuitively, termlink evidence existing for a child CV term could influence the confidence and support scores of the parent CV term. By mining the termlinks of the child and parent CV terms independently, we may be ignoring this potential contribution from the structure of the ontologies.

The second limitation is that we did not consider any patterns of annotation in a dataset of termlinks. Suppose we consider a user dataset of an OMIM record conceptually linked to a set of Entrez Gene records. Such a set of gene records have some biological affinity since they are all associated with the disease in the OMIM record. Our analysis of such sets of gene records and the corresponding datasets of termlinks indicates that patterns of annotation do exist. One such pattern is an increase in the frequency of annotation using sibling CV terms.

This research will exploit both sources of knowledge, i.e., the *is-a* structure of ontologies and the pattern of annotations. We aggregate the termlinks associated with a parent CV term, so as to use this evidence to potentially boost the values for confidence and support scores in associations of the parent CV term. A weight factor (α) determines the relative weight of evidence or the contribution from the child CV terms. The value of α can also reflect a variance of confidence scores of the sibling CV terms of some parent CV term, e.g., a high variance can reduce the contribution from child terms.

Using a background dataset of OMIM, Entrez Gene and PubMed, and user datasets that have observable patterns of annotation, we demonstrate the benefits of our research. One interesting result is that we find potentially significant associations involving parent GO terms, where this term *does not occur* among the termlinks, i.e., it was not used for annotation or it was not explicitly linked to the partner MeSH term. This suggests that our method can identify implicit annotations and there is scope to generate new knowledge as we identify such new associations.

The paper is organized as follows: Section 2 defines a background *LSLink* dataset, *termlinks* and *support and confidence*. Section 3 motivates the potential benefits of aggregation over CV terms. Section 4 presents a methodology for determining aggregate confidence and support scores. We discuss the limitations of a simple solution (*1-step link aggregation*) and the features of a more robust solution (*2-step score-score aggregation*). Section 5 presents the related work and Section 6 illustrates evaluation results. Section 7 offers our conclusions.

2   Prior Results on LSLINK Association Mining

2.1   Support and Confidence for *LSLINK* Mining

A background *LSLink* dataset is associated with a specific experiment protocol to gather a representative sample of data records, hyperlinks and annotations. Figure 2 illustrates three sample hyperlinks between two Entrez Gene and two PubMed records. The hyperlinks are between records e_1 and p_1, e_2 and p_1, and e_2 and p_2. The terms g_a, g_b, g_c and m_a, m_b, m_c, m_d annotate these records. Each record is associated with two terms. If we consider the hyperlink between e_1 and p_1, the two CV terms g_a and g_b annotating e_1, and the two CV terms m_a and m_b annotating p_1, then we can generate four *termlinks*. An example termlink is the following: (g_a, m_c, e_2, p_2) = (DNA repair, Mitosis, 675, 10749118). These three hyperlinks from Figure 2 generate twelve termlinks. Note that both hyperlinked data records must be annotated in order to generate a termlink.

The set of termlinks represents a bridge of associations between pairs of CV terms across two CVs or ontologies. We apply support and confidence metrics from association rule mining [8,9] to identify significant pairs of associations among CV terms. The metrics reflect the extent to which the association between a pair of CV terms deviates from one resulting from chance alone (a random association). Datasets and cardinalities are defined as follows:

- (G, M, E, P) is the background dataset of genes from Entrez Gene (E) annotated with GO terms (G) with links to PubMed records (P) that are in turn annotated by MeSH terms (M). Termlinks are derived from this dataset. $\#(G, M, E, P)$ is the cardinality of the termlinks in (G, M, E, P). (G, M, E', P') and $\#(G, M, E', P')$ correspond to the user dataset, a subset of the background dataset.
- $\#(g_u \wedge m_w, E, P)$ is the cardinality of termlinks containing the pair of terms g_u and m_w in the background dataset. $\#(g_u \wedge m_w, E', P')$ is the corresponding value in a user dataset.
- $\#(g_u \vee m_w, E, P)$ is the cardinality of termlinks containing either term g_u or term m_w in the background dataset. $\#(g_u \vee m_w, E', P')$ is the corresponding value in a user dataset.

Finally, we define support and confidence as follows:

$$Supp(g_u, m_w, E', P') = \frac{\#(g_u \wedge m_w, E', P')}{\#(G, M, E', P')} \qquad (1a)$$

$$Conf(g_u, m_w, E', P') = \frac{\#(g_u \wedge m_w, E', P')}{\#(g_u \vee m_w, E', P')} \qquad (1b)$$

2.2   Results from Mining

There can be a potentially large number of associations of pairs of CV terms even for a single gene. For example, for a user dataset defined for the human gene TP53 [7], there were 986,612 termlinks and they represented 83,116 distinct associations between pairs of GO and MeSH terms! The support and confidence metrics were used to rank these pairs of associations and identify the Top 25 potentially significant pairs for each gene. Experts (medical doctors and cancer researchers) rated the associations of pairs of CV terms along the following independent dimensions: (Meaningful, Maybe Meaningful, Not Meaningful), and (Widely Known, Somewhat Known, Unknown/Surprising). A majority of the Top 25 pairs of associations for user datasets such as BREAST CANCER, CFTR, TP53, etc., were identified as a true positive. Several of the pairs were unknown and might lead to new knowledge. For example, for BREAST CANCER, the previously unknown association of the GO term negative regulation of centriole replication with the MeSH term Fallopian Tube Neoplasms might be interesting, because it indicates that the tumor and the negative regulation might have a causal relationship [10]. The background dataset of termlinks from this study and the associations among pairs of GO and MeSH terms are available at the following site: http://www.cbcb.umd.edu/research/lslink/lodgui/

3 Motivation for Aggregation

We first illustrate the potential benefit of exploiting structural knowledge of *is-a* hierarchies and then discuss patterns of annotation.

The first set of examples are from termlinks generated from a user dataset of the human gene TP53 in Entrez Gene, PubMed records that are hyperlinked to it, and the corresponding annotations. Consider the GO and MeSH *is-a* hierarchies of Figure 3. In Figure 3(a), a termlink (negative regulation of progression through cell cycle, Cyclin-Dependent Kinases, 7157, 17612495) occurs between the parent GO term and the parent MeSH term. In addition, two termlinks (cell cycle arrest, CDC2-CDC28 Kinases, 7157, 14640983) and (cell cycle arrest, Cyclin-Dependent Kinase 2, 7157, 17371838) occur between the child terms. These latter two termlinks are evidence to boost the association between the pair of parent terms.

In Figure 3(b), the termlink (protein binding, Tosylphenylalanyl Chloromethyl Ketone, 7157, 12821135) occurs between the parent GO term protein binding and a child MeSH term Tosylphenylalanyl Chloromethyl Ketone. In addition, there are two termlinks from the parent MeSH term to two child GO terms. Note that *there is no termlink* between the two parent CV terms, protein binding and Amino Acid Chloromethyl Ketones in the termlink dataset; this is represented by a broken link between the pair of terms in the association bridge. However, the three termlinks in this Figure can be considered evidence to *introduce* a new association between the parent GO term protein binding and the parent MeSH term Amino Acid Chloromethyl Ketones.

To summarize, Figure 3 presented two examples of termlinks associated with combinations of parent/child CV terms. It seems intuitively apparent that the termlink evidence attached for example to the child GO terms should influence the evidence of the parent GO terms. By treating these termlinks as strictly independent, we may be ignoring potentially valuable information offered by

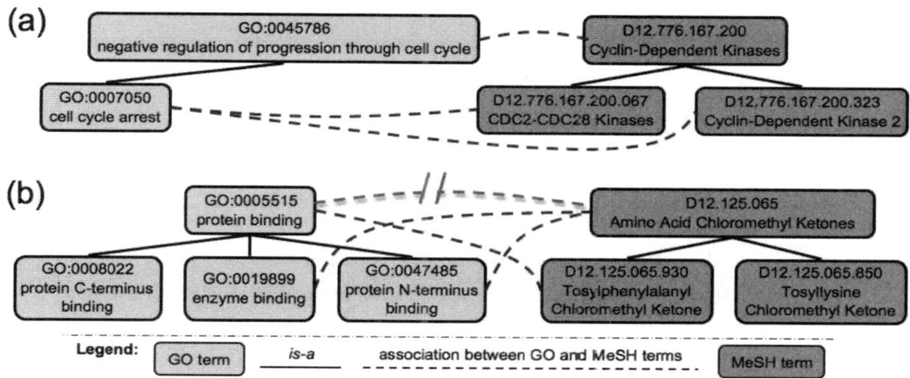

Fig. 3. Example parent-and-child hierarchies in GO and MeSH (each dotted line show an actual association generated in the human gene TP53 user dataset)

the structure of the GO ontology. Note that this applies to each participating ontology involved in generating termlink, in this case GO and MeSH. Thus, analogously from the perspective of the MeSH hierarchy, parent MeSH terms may benefit from the termlink evidence of their child MeSH terms. Finally, *new associations* between pairs of parent CV terms may also be introduced, where the parent CV term was *not used* for annotation.

Note that in the experiments reported in this paper, we only exploit a limited amount of knowledge. For example, we limit aggregation of termlink evidence along the GO *is-a* hierarchy alone, and we only consider aggregation from a GO CV term to its immediate parent term. We plan to study multiple-level aggregation along both the GO and MeSH hierarchies in future research.

Next, we illustrate a pattern of annotation that results in a higher frequency of annotations that use sibling terms from the GO ontology. We note that there is a similar pattern of higher frequency of annotation of parent and child terms, and that these patterns are also observed in individual Entrez Gene record annotations. For lack of space, we do not provide evidence on all such patterns.

We consider a dataset of termlinks obtained from OMIM records conceptually hyperlinked to (one or a set of) gene records in Entrez Gene. We note that these gene records are biologically linked since they are associated with the same disease in the OMIM record. As of September 6, 2007, there were 14,851 OMIM records. The distribution of Entrez Gene records conceptually linked to an OMIM record is given in Figure 4. While 14,502 OMIM records are linked to a single gene, 193 records have links to two genes, and SCHIZOPHRENIA (MIM Number 181500) links to 22 genes.

To illustrate the annotation pattern, we compare two techniques to group pairs of gene records to create user datasets. For the first method (**OM linked**), we place a pair of genes in a user dataset only if both genes are conceptually hyperlinked to the same OMIM record. Next, we generate a similar number of pairs for **Random**; here we pick a pair of human genes at random from

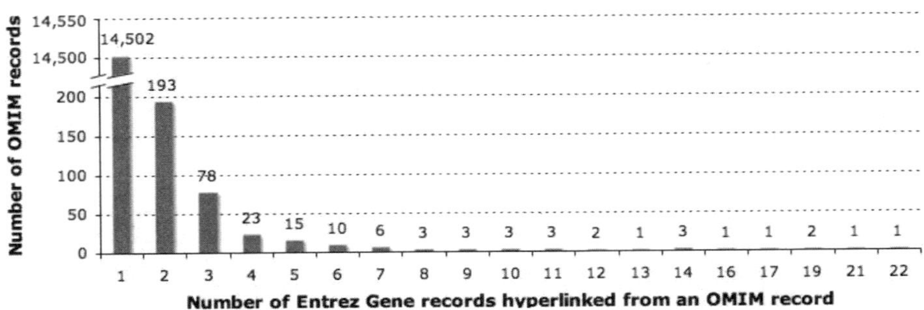

Fig. 4. Distribution of Entrez Gene records hyperlinked per OMIM record

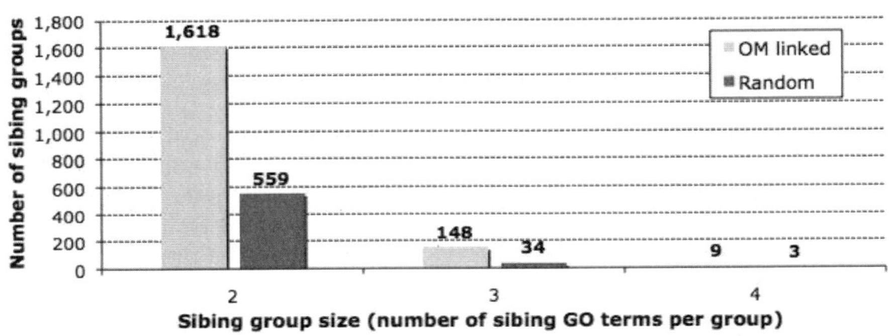

Fig. 5. Distribution of number of sibling GO terms for 1,000 pairs of genes

Entrez Gene. For each pair in **OM linked** and **Random**, we extract the GO annotations. Each dataset contains 1,000 pairs of genes. Figure 5 shows the distribution of the number of sibling GO terms that annotate the pairs of genes from **OM linked** and **Random**.

We observe that pairs of genes in **OM linked** have a *much higher distribution* of sibling GO terms. For example, there are 1,618 occurrences of (pairs of) termlinks involving a pair of sibling GO terms, and 148 occurrences of (a triple of) termlinks involving a triple of sibling GO terms, in **OM linked**. In contrast, the 1,000 pairs of genes in **Random** only have 559 occurrences of pairs and 34 occurrences of triples of sibling GO terms. To validate the pattern of annotation, we generated the 1,000 pairs of **OM linked** genes and the 1,000 pairs of **Random** genes three times. The three **OM linked** datasets had a mean of 1,499 pairs of sibling GO terms and a mean of 196 triples of GO terms. The three **Random** datasets had a mean of 487 pairs of sibling GO terms and a mean of 41 triples of GO terms. To summarize, user dataset such as **OM linked** with pairs of genes with biological affinity reflect a pattern of annotation with a higher frequency of annotation using sibling GO terms.

4   Methodology for Aggregation

We consider boosting the support and confidence scores of associations of the parent CV terms using the evidence of the termlinks of child CV terms. We use the unboosted score for support or confidence in Equations 1a and 1b as a baseline, $Supp_B$ or $Conf_B$, respectively.

We propose two solutions for aggregation. The simple solution, **1**-step **L**ink aggregation (1L), will aggregate the termlinks from the child to the parent and use a counting approach. This approach has two limitations. One is that the percentage contribution from the termlinks of the child CV term cannot be controlled. The second is that a variance of confidence among the sibling terms of the parent CV term cannot be factored in by the 1L simple counting approach. We then present a comprehensive solution, **2**-step **S**core-**S**core (2SS), that obtains a weighted score for the parent CV term. The weighted score allows the contribution from the child CV terms to be controlled. The value of the weight α can reflect the variance of confidence of the sibling CV terms. For example, a high variance can increase the contribution from the child terms.

4.1   Simple Solution for Aggregation (1L)

Consider the example in Figure 6(a) where g_1 and g_2 are two sibling child terms of parent GO term g_u. There are 2 termlinks, one from GO term g_u, and another one from g_2, to the MeSH term m_w. The confidence scores for the parent g_u, or for the child g_2, paired with m_w, are $\frac{1}{4}$ and $\frac{1}{3}$, respectively.

The *simple* 1L counting based approach to boost the confidence score of the parent CV term g_u will accumulate all termlinks associated with g_2 and credit it to the parent term. The 1L expression for the boosted support and confidence scores for the parent term is as follows:

$$Supp_{1L}(g, m, E', P') = \frac{\#(g \wedge m, E', P') + \#(g_i \wedge m, E', P' | g_i \in Child(g))}{\#(G, M, E', P')}$$

(2a)

$$Conf_{1L}(g, m, E', P') = \frac{\#(g \wedge m, E', P') + \#(g_i \wedge m, E', P' | g_i \in Child(g))}{\#((g \vee g_i) \vee m, E', P' | g_i \in Child(g))}$$

(2b)

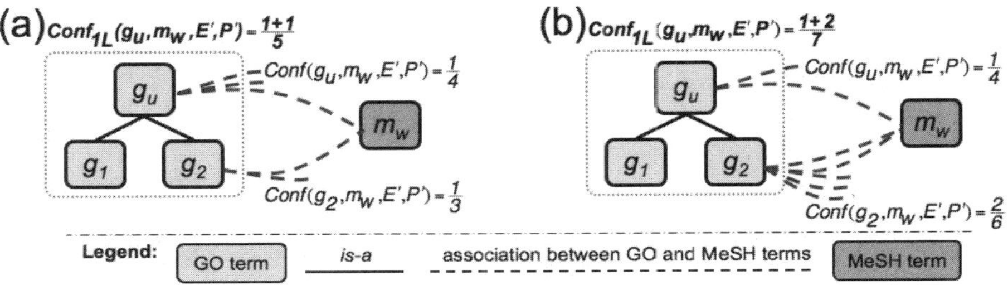

Fig. 6. Examples of one-level *1L Aggregation* from child terms to parent term

In this example, the original confidence for the association between parent g_u and m_w was $\frac{1}{4}$, and the boosted confidence score is $\frac{1+1}{5} = \frac{2}{5}$.

4.2   Limitations of the Simple Solution

We present two cases that illustrate the limitation of the simple 1L counting approach. Consider the termlinks of Figure 6(b). The original confidence scores for the associations of g_u, and g_2, with m_w, are $\frac{1}{4}$ and $\frac{2}{6}$, respectively. We note that these values are equal to the scores in Figure 6(a). Suppose that we use the simple counting 1L approach to boost the confidence score. The boosted value for confidence for the association between g_u and m_w will be $\frac{1+2}{7} = \frac{3}{7}$.

We note that the boosted confidence score of $\frac{3}{7}$ in Figure 6(b) between g_u and m_w is different from the boosted value of $\frac{2}{5}$ of Figure 6(a). However, in both cases, the original confidence scores between g_u and m_w, and between g_2 and m_w, are identical. This is the first limitation. Intuitively, we would like to control the contribution made by termlinks from the child CV terms, so that in a case such as Figures 6(a) and (b), when the confidence of the child CV term is the same, then there is an identical contribution to the parent CV term. With the 1L approach, the contribution to the parent CV term is not controlled by the confidence of the child CV term but instead it is controlled by the number of termlinks that refer to the child CV terms.

We next consider the situation where there is a variance in the confidence of the associations of the sibling CV terms. In Figure 7(a), the confidence scores for the associations of each of child terms, g_1 or g_2, with m_w, is $\frac{3}{8}$, i.e., they are of equal confidence. In Figure 7(b), there is a *variance* of the confidence scores of the child terms. The confidence score of the association of g_1 with m_w is $\frac{1}{8}$, while the confidence score in the association of g_2 with m_w is 5 times higher and is $\frac{5}{8}$.

In both Figures 7(a) and (b), the original confidence score of the association of the parent g_u with m_w is $\frac{1}{8}$. Using the 1L approach, the boosted confidence score for the association between g_u and m_w is also $\frac{1+3+3}{10} = \frac{7}{10}$, in both cases. Ideally, when there is equal confidence in the associations of the sibling terms (as in Figure 7(a)), this may be considered strong evidence that these siblings

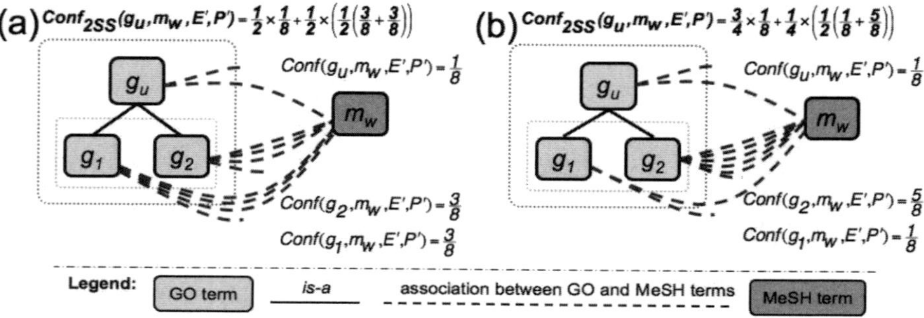

Fig. 7. Examples of one-level *2SS Aggregation* from child terms to parent term

should boost the confidence in the associations of the parent term. On the other hand, when there is a significant variance in the confidence of the sibling terms (as in Figure 7(b)), it is unclear if these siblings are providing strong evidence to boost the confidence in the parent term. Thus, referring to Figures 7(a) and (b), when there is no variance in the confidence scores of the siblings as in Figure 7(a), the boost to the parent should be greater.

4.3   Comprehensive Solution for Aggregation (2SS)

We present the 2SS aggregation method; it will overcome both limitations of the 1L approach. It will use a weight factor α to control the contribution to the parent CV term using the confidence of the child CV terms. The value of α will be determined based on the variance of the confidence of the sibling CV terms. The support and confidence scores presented in Equations (3a) and (3b).

$$Supp_{2SS}(g, m, E', P')$$
$$= (1 - \alpha) * Supp(g, m, E', P') + \alpha * Avg(Supp(g_i, m, E', P')|g_i \in Child(g))$$
$$\tag{3a}$$

$$Conf_{2SS}(g, m, E', P')$$
$$= (1 - \alpha) * Conf(g, m, E', P') + \alpha * Avg(Conf(g_i, m, E', P')|g_i \in Child(g))$$
$$\tag{3b}$$

We summarize the features of the 2SS solution. First, we calculate the confidence score for each of the child terms, and then we average the confidence scores over all the child terms. We then use a weighting factor α to determine the actual contribution from the the child terms that should be used to boost the confidence score of the parent. We experiment with the following simple rule-of-thumb to determine a value for α between 0 and $\frac{1}{2}$, where the value for α will depend on the *variance* in the confidence scores for the child terms. To explain, if there is *high* variance in the confidence score for each of the child terms of some parent g_u, then we will be *less confident* that we should aggregate over these child terms and use the child terms to potentially boost the confidence score in g_u. If the variance in the confidence scores for the child terms is *low*, we assign $\alpha = \frac{1}{2}$ to show that there is an equal importance between the weight given to the parent term and the weight given to the child terms.

We note that based on the above expression, the boost to the parent g_u is greatest when the confidence score of each of the child terms is independently high, and when there is low variance in the confidence score of the child terms. The boost to g_u is low when either the confidence score in each of the child terms is low, or when there is a high variance in the confidence scores of all child terms of g_u. The boosted confidence score (with $\alpha = \frac{1}{2}$) in Figure 7(a) is $\frac{1}{2} \times \frac{1}{8} + \frac{1}{2} \times (\frac{3}{8} + \frac{3}{8}) = \frac{1}{4}$. This value is higher compared to the boosted confidence score (with $\alpha = \frac{1}{4}$) in Figure 7(b) which is $\frac{3}{4} \times \frac{1}{8} + \frac{1}{4} \times (\frac{1}{8} + \frac{5}{8}) = \frac{3}{16}$. Although the difference between these two boosted confidence values is $\frac{1}{16}$, this difference can have a *major* impact on the rank of the associations. However, in our experiments, we use the same value of α for all associations.

We note that the rule of thumb used to select a value of α will need to be expanded to consider aggregation along multiple levels of the GO hierarchy, as well as simultaneous aggregation along both the GO and MeSH hierarchies.

5   Related Work

We consider related work in generalized association rule mining, ranking and ontology matching. Generalized association rule mining [11][12][13] creates an extended transaction set either by *replacing* an item with a new item representing a generalized concept, or by *aggregating* both the original item and the generalized item. We note that the generalized concept does not occur in their original transaction set. Their solution approach is similar to our counting based 1L approach and faces the limitations that were discussed, i.e., controlling the contribution of child CV terms and reflecting variance of confidence. [14] proposed to assign a lower threshold of *support* for associations in the lower levels of ontology. Furthermore, in order to reduce the search space by filtering associations contain independent items, the metric *usefulness* or *interest* is suggested by [15]. They define *R-interesting* as a rule is interesting iff it has no predecessor or its adjacent interesting predecessor is *R-interest*[16].

While there is extensive literature on ranking using link structure of a graph, the focus is on ranking nodes in a general graph [17,18,19]. There is no work on ranking an association bridge (edges) of a bi-partite graph and ranking typically does not consider metadata such as the *is-a* hierarchy.

There is also research on ontology matching or ontology alignment [20,21,22]. The objective is to determine matches or correspondence between concepts or between subgraphs. Their solutions are based on string similarity between the labels of concepts, structural similarity and relationship patterns in the ontology. [22] uses a technique similar to association rule mining. While this research exploits similar knowledge, since the objectives are different, we typically cannot apply any of their solutions.

6   Experimental Evaluation

6.1   Generating User Datasets

Disease related user datasets were generated using the corresponding OMIM record. The protocol follows links from OMIM to Entrez Gene and then to PubMed. Table 1 reports on the statistics of four disease related datasets. For e.g., for the **BREAST CANCER** user dataset, the OMIM record has hyperlinks to 13 Entrez Gene records that are annotated with 147 distinct GO terms. Following the hyperlinks from these 13 Entrez Gene records to PubMed, we obtain 3,237 distinct PubMed records that are annotated with 2,463 distinct MeSH descriptor terms (of selected UMLS semantic types [23]). We generate 1,232,086 termlink instances and collect 124,342 distinct associations pairs of a GO term and a MeSH term. The one-level aggregation using the GO structured *is-a* hierarchy

Table 1. Statistics in four disease-related user datasets

MIM Number Title	114480 breast cancer	114500 colorectal cancer	176807 prostate cancer	191170 tumor protein p53
$\#(E')^1$	13	14	13	1^2
$\#(G)^3$	147	135	117	44
$\#(P')^4$	3,237	2,827	1,518	1,888
$\#(M)^5$	2,463	2,594	1,624	1,889
$\#(G, M, E', P')$	1,232,086	1,189,379	339,491	986,612
$\#(G, M)^6$	124,342	123,343	57,735	83,116
$\#(G_{new})^7$	24	23	20	7
$\#(G_{new}, M)^8$	18,648	18,002	9,539	13,223

introduces 24 new GO terms and 18,648 pairs of associations that did not occur
among the original termlinks.

6.2 Examples of Identifying Significant Associations Via Aggregation

We use several user datasets to illustrate a range of opportunities to boost the
associations of the parent CV terms. We note that all these examples have been
verified to be meaningful and some are previously unknown.

We calculate a baseline confidence score, $Conf$, for associations of the parent
CV term that does not reflect aggregation evidence, and a boosted confidence
score $Conf_{2SS}$. We also report on the original rank $Rank$ and the new rank
$Rank_{2SS}$. Note that for each user dataset, $Rank_{2SS}$ is determined over a combination (union) of both the original pairs of associations of CV terms and
any new associations introduced via aggregation. For example, for the BREAST
CANCER dataset, $Rank_{SS}$ will be determined over (124,342+18,648) associations.
We use constant values of $\alpha = \frac{1}{2}$ in the following four examples. Please note that
the boosted ranks on the child terms can be worsen than the baseline ranks, because the newly introduced parent term may have better ranks and the ranks of
some other parent terms may have been improved more.

The first example in Table 2 involves a parent GO term DNA binding and
its three child terms, transcription factor activity, damaged DNA binding
and sequence-specific DNA binding. The associated MeSH term is Cell
Cycle Proteins. We see that the parent term already has the highest confidence score (among these associations) and has a rank of 156. The confidence

---

[1] $\#(E')$: number of Entrez Gene records hyperlinked to the OMIM record.
[2] Corresponding to the human gene TP53 dataset in Sections 2 and 3.
[3] $\#(G)$: number of distinct GO terms annotating E'.
[4] $\#(P')$: number of distinct PubMed records hyperlinked to E'.
[5] $\#(M)$: number of distinct MeSH terms annotating P'.
[6] $\#(G, M)$: number of distinct CV term associations.
[7] $\#(G_{new})$: number of new GO terms introduced by aggregation.
[8] $\#(G_{new}, M)$: number of distinct CV term associations generated by aggregation.

Table 2. BREAST CANCER user dataset having MeSH descriptor term Cell Cycle Proteins

GO Term	Parent GO Term	$Conf$	$Rank$	$Conf_{2SS}$	$Rank_{2SS}$
DNA binding		0.0180	156	0.0099	**133**
transcription factor activity	DNA binding	0.0045	2,572		3,522
damaged DNA binding	DNA binding	0.0005	31,030		38,349
sequence-specific DNA binding	DNA binding	0.0005	31,030		38,349

Table 3. BREAST CANCER user dataset having MeSH descriptor term 1-Phosphatidylinositol 3-Kinase

GO Term	Parent GO Term	$Conf$	$Rank$	$Conf_{2SS}$	$Rank_{2SS}$
phosphoinositide 3-kinase activity				0.0125	**71**
phosphatidylinositol-4,5-bisphosphate 3-kinase activity	phosphoinositide 3-kinase activity	0.0325	29		32
1-phosphatidylinositol-3-kinase activity	phosphoinositide 3-kinase activity	0.0175	161		195

score of the child terms are low and they are farther back in rank. There is also high variance in the confidence score of the child terms. Nevertheless, there is a positive contribution from the child terms and the parent term's boosted rank is 133. We note that the actual confidence score of the parent term has gone down after boosting and we note that in general the scores for confidence score tend to reduce after boosting. However, the rank is determined using the score relative to other associations. Thus, while the actual score may reduce, the rank may actually be improved.

In the second example in Table 3, the parent term phosphoinositide 3-kinase activity does not have a confidence score since there are no termlinks for this GO term to the MeSH term 1-Phosphatidylinositol 3-Kinase. The parent term has two child terms, phosphatidylinositol-4,5-bisphosphate 3-kinase activity and 1-phosphatidylinositol-3-kinase activity. Both child terms have high confidence scores and their ranks are also very good, at 29 and 161, respectively. The variance in the child terms is also low. This is a situation where the boost provided by the child terms should be the most significant, i.e., the confidence score in the child terms is high and variance in confidence score is low. Thus, after the parent term is boosted, it too has a very good rank of 71. We note that the rank of the child terms terms has worsened slightly. To explain, there are several parent GO term associations that did not occur in the original termlinks that have been introduced after aggregation. They tended to be ranked ahead of the child terms from the example.

In the third example in Table 4, we consider the parent GO term protein binding in the COLORECTAL CANCER user dataset. The parent GO term has four child terms, enzyme binding, protein N-terminus binding, protein

Table 4. COLORECTAL CANCER user dataset having MeSH descriptor term Tumor Suppressor Protein p53

GO Term	Parent GO Term	$Conf$	$Rank$	$Conf_{2SS}$	$Rank_{2SS}$
protein binding		0.0165	147	0.0126	**93**
enzyme binding	protein binding	0.0174	101		129
protein N-terminus binding	protein binding	0.0174	101		132
protein C-terminus binding	protein binding	0.0004	40,481		47,729
insulin receptor substrate binding	protein binding	0.0001	117,248		133,069

Table 5. PROSTATE CANCER user dataset having MeSH descriptor term Kangai-1 Protein

GO Term	Parent GO Term	$Conf$	$Rank$	$Conf_{2SS}$	$Rank_{2SS}$
integral to membrane		0.0429	14	0.0394	1
integral to plasma membrane	integral to membrane	0.0360	26		30

C-terminus binding and insulin receptor substrate binding. The confidence scores of the associations of child terms enzyme binding and protein N-terminus binding is high and their rank is 101. The confidence scores of the other two child terms is very low. This is a case where the confidence scores in two child terms is high and there is also high variance among the child terms' confidence scores. The boost should not be as significant as in the previous case. We see that the parent rank has improved from 147 to 93. Thus, the boost is not as significant as in Table 3.

In the final example in Table 5, we consider the PROSTATE CANCER user dataset. The parent term integral to membrane has only one child term integral to plasma membrane. The associated MeSH term is Kangai-1 Protein. Both parent and child have high confidence scores and their rank is within the Top 30. The boosted confidence score for the parent term pushes it to rank *first* among the (57,735+9,539) associations for this user dataset! To summarize, we use a variety of GO *is-a* hierarchies, and range of confidence scores for the child terms, to illustrate the impact on the parent CV term.

6.3   Impact of α on Boosted Rank

We consider the BREAST CANCER dataset; it has 124,342 associations prior to aggregation and 18,642 associations are added after aggregation. We select the Top 300 associations (after 2SS boosting). Figure 8 reports on the rank $Rank$ before boosting (Y axis) and the rank $Rank_{2SS}$ after boosting (X axis), for the Top 300. If an association did not occur in the original termlink dataset, its rank is labeled no rank on the Y axis. We compare two α values, $\frac{1}{2}$ and $\frac{1}{4}$.

A 45 degree line in Figure 8 represents the case where there is no change in the rank from boosting. For $\alpha=\frac{1}{4}$ (labeled +), the contribution from the child

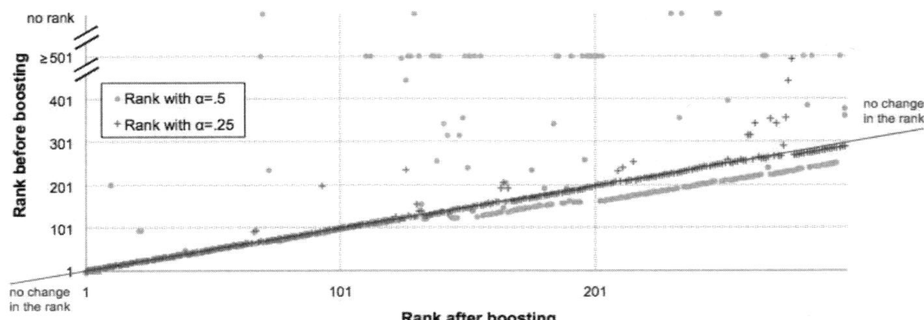

Fig. 8. Impact (rank changes) of boosting confidence scores for **BREAST CANCER** user dataset

terms is only 25%; hence we see many of these datum clustered around the **no change in rank** line. There are a few datum scattered above the line indicating cases where the ranks have improved after boosting.

For $\alpha=\frac{1}{2}$ (labeled •), the situation is quite different since the contribution from the child terms is more significant at 50%. Many of the datum above the baseline indicate improvement of the rank. Among these improvements, there are six new associations (originally with **no rank**) and 21 associations whose original ranks were greater than 8,000 that now occur in the Top 300.

7   Conclusion

We have presented an approach and preliminary evaluation to exploit knowledge from ontologies and patterns of annotation to identify significant associations jointly offer a bridge between a pair of ontologies. In future work, we will consider further extensions, e.g., aggregating simultaneously using the structure of both ontologies, aggregating up multiple levels, etc. We also plan an extensive evaluation on termlinks to identify interesting patterns of annotation, and study their impact on finding significant associations.

Acknowledgments. This research has been partially supported by the National Science Foundation under grants IIS0222847, IIS0430915 and 0312356.

References

1. Maglott, D.R., Ostell, J., Pruitt, K.D., Tatusova, T.: Entrez Gene: gene-centered information at NCBI. Nucleic Acids Research 35, D26–D31 (2007) (Database issue)
2. Hamosh, A., Scott, A.F., Amberger, J.S., Bocchini, C.A., McKusick, V.A.: Online Mendelian Inheritance in Man (OMIM), a knowledgebase of human genes and genetic disorders. Nucleic Acids Research 33, D514–D517 (2005) (Database issue)

3. Wheeler, D.L., Barrett, T., Benson, D.A., Bryant, S.H., Canese, K., Chetvernin, V., Church, D.M., DiCuccio, M., Edgar, R., Federhen, S., Feolo, M., Geer, L.Y., Helmberg, W., Kapustin, Y., Khovayko, O., Landsman, D., Lipman, D.J., Madden, T.L., Maglott, D.R., Miller, V., Ostell, J., Pruitt, K.D., Schuler, G.D., Shumway, M., Sequeira, E., Sherry, S.T., Sirotkin, K., Souvorov, A., Starchenko, G., Tatusov, R.L., Tatusova, T.A., Wagner, L., Yaschenko, E.: Database resources of the National Center for Biotechnology Information. Nucleic Acids Research 36, D13–D21 (2008) (Database issue)
4. Wang, A.Y., Sable, J.H., Spackman, K.A.: The SNOMED Clinical Terms development process: refinement and analysis of content. In: AMIA 2002 Annual Symposium, San Antonio, Texas, USA, November 9-13, 2002, pp. 845–849 (2002)
5. Gene Ontology Consortium: The Gene Ontology (GO) project in 2006. Nucleic Acids Research 34, 322–326 (2006) (Database issue)
6. Savage, A.: Changes in MeSH data structure. Technical Report (313), NLM Technical Bulletin (March-April 2000)
7. Lee, W.J., Raschid, L., Srinivasan, P., Shah, N., Rubin, D., Noy, N.: Using annotations from controlled vocabularies to find meaningful associations. In: Cohen-Boulakia, S., Tannen, V. (eds.) DILS 2007. LNCS (LNBI), vol. 4544, pp. 247–263. Springer, Heidelberg (2007)
8. Agrawal, R., Imielinski, T., Swami, A.: Mining association rules between sets of items in large databases. SIGMOD Record 22(2), 207–216 (1993)
9. Agrawal, R., Srikant, R.: Fast Algorithms for Mining Association Rules in Large Databases. In: Proceeding of the 20th International Conference on Very Large Data Bases (VLDB 1994), San Francisco, CA, USA, September 1994, pp. 487–499 (1994)
10. Day, C.P.: Personal communiction (2007)
11. Tseng, M.C., Lin, W.Y., Jeng, R.: Incremental maintenance of ontology-exploiting association rules. In: International Conference on Machine Learning and Cybernetics, Hong Kong, China, August 19-22, 2007, pp. 2280–2285 (2007)
12. Han, J., Fu, Y.: Discovery of multiple-level association rules from large databases. In: Proceeding of the 21th International Conference on Very Large Data Bases (VLDB 1995), Zürich, Switzerland, September 11-15, 1995, pp. 420–431 (1995)
13. Jiang, T., Tan, A.H., Wang, K.: Mining generalized associations of semantic relations from textual Web content. IEEE Transactions on Knowledge and Data Engineering 19(2), 164–179 (2007)
14. Cheung, D.W.L., Ng, V.T.Y., Tam, B.W.: Maintenance of discovered knowledge: a case in multi-level association rules. In: Second International Conference on Knowledge Discovery and Data Mining (KDD 1996), Portland, Oregon, USA, pp. 307–310 (1996)
15. Srikant, R., Agrawal, R.: Mining generalized association rules. Future Generation Computer Systems 13(2-3), 161–180 (1997)
16. Wang, X., Ni, Z., Cao, H.: Research on association rules mining based-on ontology in e-commerce. In: International Conference on Wireless Communications, Networking and Mobile Computing (WiCOM 2007), Shanghai, China, September 2007, pp. 3544–3547 (2007)
17. Brin, S., Page, L.: The anatomy of a large-scale hypertextual Web search engine. Computer Networks and ISDN Systems 30(1–7), 107–117 (1998)
18. Page, L., Brin, S., Motwani, R., Winograd, T.: The pagerank citation ranking: Bringing order to the web. Technical report, Stanford Digital Library Technologies Project (1998)

19. Kleinberg, J.M.: Authoritative sources in a hyperlinked environment. Journal of the ACM 46(5), 604–632 (1999)
20. Rahm, E., Bernstein, P.A.: A survey of approaches to automatic schema matching. VLDB Journal: Very Large Data Bases 10(4), 334–350 (2001)
21. Hopcroft, J.E., Karp, R.M.: An $n^{5/2}$ algorithm for maximum matchings in bipartite graphs. SIAM J. Comput. 2(4), 225–231 (1973)
22. Yu, C., Zavaljevski, N., Desai, V., Johnson, S., Stevens, F.J., Reifman, J.: The development of PIPA: an integrated and automated pipeline for genome-wide protein function annotation. BMC Bioinformatics 9(52) (January 2008)
23. Bodenreider, O.: The Unified Medical Language System (UMLS): integrating biomedical terminology. Nucleic Acids Research 32, D267–D270 (2004) (Database issue)

Automatic Methods for Integrating Biomedical Data Sources in a Mediator-Based System

Fleur Mougin[1,2], Anita Burgun[1], Olivier Bodenreider[3], Julie Chabalier[1],
Olivier Loréal[4], and Pierre Le Beux[1]

[1] EA 3888, IFR 140, Faculté de Médecine, University of Rennes 1, France
[2] LESIM, INSERM U593, ISPED, University of Bordeaux 2, France
[3] National Library of Medicine, Bethesda, Maryland, USA
[4] INSERM U522, IFR 140, University of Rennes 1, CHU Pontchaillou, France
`fleur.mougin@isped.u-bordeaux2.fr`
`{anita.burgun,julie.chabalier,olivier.loreal,`
`pierre.le-beux}@univ-rennes1.fr, olivier@nlm.nih.gov`

Abstract. The information needed by biologists and physicians for research purposes is distributed over many heterogeneous sources. Integration systems provide a single, centralized and homogeneous interface for users to query multiple information sources simultaneously. The major limitation of integration systems, including mediator-based systems, is that the tasks involved in their creation and maintenance remain mainly manual. To address this limitation, we developed automated methods for facilitating the creation of a mediator-based system. We first implemented an automatic method for acquiring the local schemas of the sources to be integrated. We derived the global schema from the UMLS. Finally, we proposed *schema-* and *instance-*based approaches to mapping data elements from the local schemas to the global schema. To illustrate the applicability of our methods, we created a mediator-based system integrating eleven biomedical sources. This prototype is operational, available on the Internet (http://www.med.univ-rennes1.fr/cgi-bin/mougin/These/system.pl) and its evolution is managed semi-automatically.

Keywords: data integration, mediator-based approach, schema-level mapping methods, instance-level mapping methods, biomedicine.

1 Introduction

Most of the information needed by physicians and biologists for research purposes is present in electronic biomedical resources available through the Internet. In addition, the biomedical domain is in constant evolution and generates considerable amounts of data. Collecting information manually is thus slow and error-prone. Integrating biomedical sources in order to facilitate global access to multiple, heterogeneous sources has become unavoidable [1]. Moreover, an integration system adapted to the biomedical domain should be easy to use for biologists and physicians, scalable, and provide up-to-date information.

Three main integration approaches have been proposed to reconcile distributed sources in the biomedical domain:

A. Bairoch, S. Cohen-Boulakia, and C. Froidevaux (Eds.): DILS 2008, LNBI 5109, pp. 61–76, 2008.

- in *datawarehouses*, e.g., GUS [2], data are imported from various sources and stored locally in a single format. A direct limitation of datawarehouses is that, unless the local version of the sources is updated regularly in the warehouse, query results are not necessarily up-to-date. The evolution of such systems is typically a difficult issue.
- *path-based* (or navigational) *systems*, e.g., BioGuide [3], correspond to graphs in which the various entities are linked by paths, making it possible for users to navigate between sources. With such systems, users are responsible for following the links created across resources, which constitutes a limitation of navigational systems. Additionally, changes to the sources require links to be recomputed over the whole system. Unlike other approaches, path-based systems do not impose a consistent view on the sources, which greatly facilitates their evolution.
- with *mediator-based systems*, e.g., TAMBIS [4], data sources are queried dynamically. This approach guarantees that users access up-to-date information, because only the schemas of the sources (or local schemas) are stored in the system. For this reason, mediator-based systems tend to evolve gracefully. This approach also facilitates the query task, since users interact with a single unified schema, the global schema.

Existing mediator-based systems have been mostly created manually, which remains an important limitation to their scalability and automatic evolution. It is thus essential to automate the tasks involved in the creation and maintenance of such systems [5]. Practically, as shown in Fig. 1, this means automating the acquisition of local schemas (step 1), the definition of the global schema (step 2), and the mapping of the local schemas to the global schema (step 3).

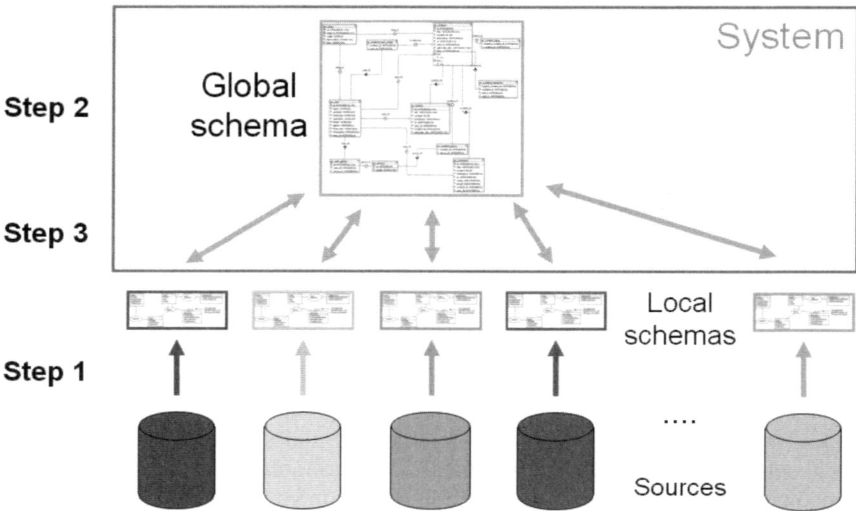

Fig. 1. The mediator-based architecture and the three major steps for its conception: 1) sources schema acquisition, 2) definition of the global schema, and 3) mapping of the local schemas to the global schema

This paper addresses automation in the creation and maintenance of systems integrating biomedical sources. More specifically, we propose automated methods for creating and maintaining mediator-based systems and apply them to a system we developed for integrating biomedical sources accessible over the Internet. The rest of this paper is organized as follows. We first present a method for extracting local schemas, based on the parsing of their Web pages. We show how we adapt an existing biomedical resource for creating the global schema: the Unified Medical language System® (UMLS®). Then, we present two complementary approaches to mapping local schemas to the global schema of our system automatically. The first one operates directly on the data elements (attributes such as *gene symbol)*, while the other analyzes the data themselves (values such as *BRCA1)*. Finally, we present an application of these methods and examine their contribution to scalability management.

2   Materials and Methods

2.1   Materials

Biomedical data sources. In collaboration with biologists, we defined criteria for selecting biomedical data sources. To be integrated in our system, they should:

- contain data about general biomedical entities, such as genes, proteins, and diseases;
- be complementary: general and specialized data sources have to be integrated;
- be accessible over the Internet.

Among the data sources frequently used by biologists, and based on these criteria, we selected the following eleven biomedical sources for integration in our system:

- genomic sources: GeneCards[1], Entrez Gene[2], Geneloc[3], HGNC[4], HGMD[5], and MGI[6];
- protein sources: Swiss-Prot[7], PDB[8], HPRD[9], Interpro[10];
- medical sources: OMIM[11].

The UMLS. The Unified Medical Language System® (UMLS®) [6] provides the core set of concepts and relations for the global schema. The UMLS is a terminological resource that provides a wide coverage of the biomedical domain, including terminologies for specialized clinical disciplines, the biomedical literature, and genome

---

[1] http://bioinformatics.weizmann.ac.il/cards/
[2] http://www.ncbi.nlm.nih.gov/entrez/query.fcgi?db=gene
[3] http://genecards.weizmann.ac.il/geneloc/
[4] http://www.gene.ucl.ac.uk/nomenclature/
[5] http://www.hgmd.org/
[6] http://www.informatics.jax.org/
[7] http://www.expasy.org/sprot/
[8] http://www.rcsb.org/pdb/
[9] http://www.hprd.org/
[10] http://www.ebi.ac.uk/interpro/
[11] http://www.ncbi.nlm.nih.gov/entrez/query.fcgi?db=OMIM

annotation. The UMLS consists of three major components. The UMLS Metathesaurus® is assembled by integrating more than 100 sources vocabularies. It contains about 1.4 million concepts (clusters of synonymous terms) and more than 22 million relations among these concepts. The UMLS Semantic Network is a limited network of 135 semantic types. Each Metathesaurus concept is assigned to at least one semantic type. Finally, the Lexical Resources comprise the SPECIALIST Lexicon and Lexical Tools [7]. The UMLSKS API also provides various methods for identifying Metathesaurus concepts from input terms (exact and normalized matches). Additionally, the MetaMap Transfer (MMTx) program maps text to concepts in the Metathesaurus with additional flexibility (approximate match) [8].

2.2 Methods

Step 1: Acquiring Local Schemas. One major problem with biomedical sources is that their schema is often unavailable and rarely exploitable in its original form. Our aim is to develop an automatic method for acquiring the local schema of any source accessible over the Internet. No standard has been defined for creating biomedical local schemas in a uniform way. Consequently, the exploitation of existing schemas (e.g., NCBI schemas) would have required the development of a specific program for each schema. Instead, we proposed to acquire the schema of each source dynamically by extracting data elements from Web pages for each biomedical source. Data elements (DEs) can be defined as a basic unit of information, built on standard structures and having both a unique meaning and distinct units or values[12]. In database parlance, DEs correspond to attributes, while their associated values are instances. We then developed a method for typing DEs in order to make their semantics explicit.

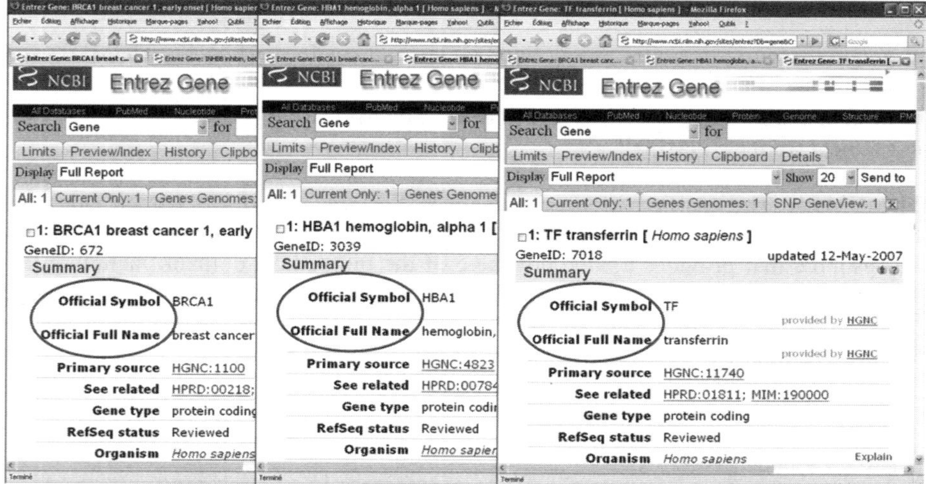

Fig. 2. Web Pages obtained by querying Entrez Gene for human BRCA1, HBA1, and TF genes. DEs correspond to invariant elements across Web pages. Examples of DEs are circled.

---

[12] http://www.atis.org/tg2k/_data_element.html

DE Extraction

Starting from a list of 100 gene names and symbols randomly extracted from the Web site of the Genetics Home Reference[13], we queried each source dynamically resulting in 100 Web pages sharing the same structure. The elements common to at least 75% of the Web pages were extracted automatically [9]. This selection resulted in eliminating specific information (e.g., a given gene name), while keeping general information (e.g., the term "Gene Name"). An example of DE extracted from the source Entrez Gene is given in Fig. 2. For instance, the terms "Official Symbol" and "Official Full Name" appear on all three pages and are therefore identified as candidate DEs.

DE Typing

We also recovered the values associated with each DE. In order to elicit the semantics of a given DE, we mapped its values to the UMLS, using exact and normalized matches (see section 2.1). We then selected the semantic type categorizing the majority of the concepts associated with a given set of values. For example, we were able to determine that the DE Official Full Name relates to gene names, because the majority of its values are categorized by the semantic type *Gene or Genome* (Fig. 3 (a)). When the type of a DE could not be determined by this process, we attempted to assign coarser predefined types. We first isolated DEs containing specific terms. For instance, when the terms "ID(s)" or "identifier" were found, the corresponding DE was typed as *Identifier*. Then, we analyzed the values characterwise and assigned the type *Sequence* to the DE when each of its non-empty values was a series of "A", "G", "C", and "T". Finally, the remaining DEs were typed as *Integer* or *String* according to their values. An example of the exploitation of DE values through heuristics is shown in Fig. 3 (b).

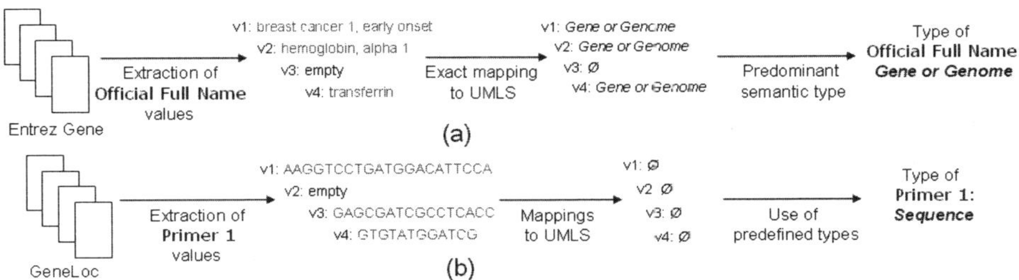

Fig. 3. Examples illustrating the typing process of DEs. (a) Official Full Name is typed through the semantic type *Gene or Genome*; (b) Primer 1 is typed as a *Sequence*.

As advocated in [10], sources schemas include general information (i.e., the name and URL of the source, as well as the kind of data in the source, e.g., gene, protein, or disease) in addition to the DEs and their types. Schemas are represented in XML, as no reasoning or specific advanced functionalities are required at this level.

Step 2: Defining the Global Schema. As mentioned earlier, our global schema was derived from the UMLS. Inappropriate links causing cycles in the UMLS hierarchies

---

[13] http://ghr.nlm.nih.gov/

were eliminated, as advocated in [11]. After transformation into a Directed Acyclic Graph, the UMLS was represented with OWL DL, one version of the Web Ontology Language often used to represent biomedical ontologies. The UMLS elements useful to our system are represented as follows:

- Semantic types and concepts are represented as classes;
- The categorization relationship between concepts and semantic types is represented as a subclass relationship;
- Hierarchical relations among semantic types were represented with the subclass relationship, as were hierarchical relations among concepts;
- Semantic types and concepts have unique identifiers (from the UMLS). Other properties include a label (the preferred term of concepts and the name of semantic types), and a textual definition, when available. Specific to concepts is the property *has_synonyms*, which contains the synonyms of the concept.

In the global schema, we represented only those UMLS semantic types, Metathesaurus concepts and relations necessary for the description of the DEs extracted from the eleven sources.

Step 3: Mapping Local Schemas to the Global Schema. This mapping aims at identifying correspondences between DEs extracted from the sources (and their values) with the concepts from the global schema, which corresponds to the notion of "schema mapping" defined in [12] and [13]. Two distinct approaches were developed for this mapping. The first one operates at the *schema* level, as it only exploits the DEs. In contrast, the second approach is based on the values associated with the DEs and lies at the *instance* level.

Schema-Level Mapping

For mapping DEs directly to the UMLS, we first attempted to find an exact match. If none was found, a match was performed after normalization. These two steps were implemented through the corresponding methods of the UMLSKS API. Finally, an approximate match was attempted using MMTx (strict model). This process resulted in three types of mappings:

- **unique match,** e.g., the DE mRNA was mapped to the concept RNA, Messenger by exact match;
- **multiple matches,** e.g., the DE Interactions resulted in an exact match to two UMLS concepts: Social Interaction and Drug Interactions;
- **no match.** Some DEs were simply not mapped to any UMLS concepts, because they are not specific to the biomedical domain. Examples of such DEs include Topology, Products, and Domains.

This automatic mapping method is efficient when a unique match is found, but is insufficient in the two other cases. More precisely, multiple matches require disambiguation and a different mapping method needs to be utilized when no direct match to UMLS concepts is found. We thus developed an alternative mapping method which exploits a different external resource: WordNet (WN) [14], an online lexical database of general English. WN is organized into a hierarchy of synsets (sets of synonymous terms) and contains more than 155,000 lexical items aggregated into about 117,000 synsets. In WN, ancestors and descendants are called hypernyms and hyponyms, respectively. Our hypothesis is that general resources such as WN could provide a complementary coverage of the domain described by the DEs under investigation. By exploiting the properties of WN, we expect to improve the mapping of DEs to the UMLS in the following ways. In case of unique matches, WN would help validate the UMLS mappings. For multiple matches, WN would contribute external information, useful for disambiguating UMLS mappings. Finally, WN would help identify indirect mappings to the UMLS when no direct UMLS mapping was found.

Validating unique mappings to UMLS. If the mapping to WN was unique, we exploited the properties of the candidate synset to validate the mapping to the UMLS. Toward this end, we compared the concept and synset according to the following criteria, in this order: 1) similarity of their definitions, 2) presence of common synonyms, and 3) presence of common ancestors. For criterion 1, after eliminating stop words, we normalized the remaining words into their base forms, which we then used for identifying common words between definitions. For criteria 2 and 3, we mapped the synonyms and hypernyms of the WN synset to the UMLS through exact and normalized matches. We then compared the results to the synonyms and ancestors of concepts obtained during the direct match of DEs to the UMLS.

Disambiguating multiple mappings to UMLS. In order to disambiguate the multiple mappings of a DE to the UMLS, we mapped it to WN, resulting in one or more synsets for this DE. We then associated pairwise the UMLS concepts and WN synsets, and selected the best (concept,synset) pair using the similarity criteria described above for the validation of unique mappings.

Identifying indirect mappings to UMLS through WN. For those DEs for which no mapping to UMLS concepts was found (i.e., when the only mapping candidates are WN synsets), we tried to find an equivalent UMLS concept not from the DE itself, but from its mapping to WN. Starting from the WN synset(s) mapped to, we first attempted to map each of the synonyms in the synset(s) to the UMLS, using exact and normalized matches as before. If no synonym was mapped to UMLS, we started an equivalent mapping process from the direct hypernyms of the synset(s). The resulting concepts constitute candidates for indirect mappings of DEs to UMLS through WN.

Instance-Level Mapping. It is also possible to map DEs to the UMLS based not on their names, but on their values. Our hypothesis is that DEs sharing a large number of values are likely to correspond to the same entity and can thus be mapped to the same UMLS concept. In practice, we computed the Jaccard similarity for each (DE_1, DE_2) pair (formula (1)) defined in [15].

$$Sim_{Jaccard} = \frac{c_1c_2}{c_1 + c_2 - c_1c_2}. \tag{1}$$

where c_1 and c_2 are the cardinalities of the value sets for DE_1 and DE_2, respectively, and c_1c_2, the cardinality of their intersection. Two DEs are deemed equivalent if the similarity between their value sets is above the threshold of 0.50, determined heuristically.

3 Results

We first report the results obtained through the methods developed to support the creation of our mediator-based system. More precisely, we present the local schemas, the global schema, and the mappings between them. Then, we present the system we created for integrating eleven biomedical sources.

3.1 Basic Elements of Our Mediator-Based System

Local Schemas. Overall, we extracted 548 DEs (474 distinct) from the eleven sources, of which 62 (11.3%) could be characterized with datatypes more specific than *String*. Detailed results are given Table 1. Local schemas are available as supplementary material at: http://www.med.univ-rennes1.fr/~mougin/schemas/.

Table 1. Results obtained for typing the DEs extracted from the sources. For each type, the number of DEs is given, followed by an example of DE and some of its associated values.

Type	Number of DEs having this type	Examples of typed DEs	Examples of associated values
Semantic type	36 (6.6%)	From (*Organism*)	Rattus norvegicus, Homo sapiens
Integer	18 (3.3%)	Molecular Weight	207732, 464482
Identifier	6 (1.1%)	Accession Numbers	U14680, X71923
Sequence	2 (0.3%)	Primer 2	GAGATCGCCTCACC
String	486 (86.9%)	Bibliography	(Earliest) J:31493 Hall JM et al., "Linkage of early-onset familial breast cancer to chromosome 17q21" Science 1990;250(4988):1684-9

The Global Schema. Overall, the global schema contains the 135 UMLS semantic types and 3,542 Metathesaurus concepts. In addition to the concepts resulting from the mapping of DEs to the UMLS, we included the ancestors of these concepts in the UMLS in order to preserve the hierarchical organization of this set of concepts for navigation purposes. The global schema is available at: http://www.med.univ-rennes1.fr/~mougin/onto/schema_global_with_wn.owl.

In addition, some concepts of the global schema have been enriched with three WN properties *has_wn_definition*, *has_wn_synonyms*, and *has_wn_hypernyms*. Actually, for those concepts mapped to WN synsets (n = 106), we chose to add the properties of these synsets to the description of the corresponding concept in the global schema, as illustrated by the concept Citation in Fig. 4. This concept was mapped to the synset citation#n#3 because their definitions share similar words (criterion 1). As a result, the concept Citation, which originally has no synonyms in the UMLS, inherits the synonyms of the synset citation#n#3 in our global schema.

```
<owl:Class rdf:ID="C0552371">
   <rdfs:label>Citation</rdfs:label>
   <has_definition>An extract or quotation from or reference to an
   authoritative source</has_definition>
   <has_wn_definition>a short note recognizing a source of information or of a quoted passage
   </has_wn_definition>
   <has_wn_synonyms>citation, cite, acknowledgment, credit, reference, mention, quotation
   </has_wn_synonyms>
   <has_wn_hypernyms>note#n#6%%comment#n#2%%statement#n#1%%mesage#n#2%%
   communication#n#2%%abstraction#n#6%%abstract_entity#n#1%%entity#n#1
   </has_wn_hypernyms>
   <rdfs:subClassOf rdf:resource="#T032"/>
   <rdfs:subClassOf rdf:resource="#C1254372"/>
</owl:Class>
```

Fig. 4. Representation of the concept Citation in the global schema. The properties obtained through WN are bold-faced.

Mapping Local Schemas to the Global Schema

Schema-Level Mapping. 387 of the 474 DEs (82%) were found directly in the UMLS, including 187 unique mappings and 200 multiple mappings. Only 87 DEs were not mapped to UMLS concepts.

As illustrated in Fig. 5 (a), WN provided supporting evidence for validating 82 unique mappings of DEs to UMLS (43.9%). WN also contributed to the disambiguation of 95 of multiple mappings (Fig. 5 (b)). Finally, 36 additional DEs were mapped to the UMLS using WN, through synonyms (16) and direct hypernyms (20), as shown in Fig. 5 (c) and (d), respectively.

Overall, 423 DEs were mapped to the UMLS and 74% of all mappings were exploitable automatically. The remaining mappings required some degree of manual intervention before they could be used in the system, including disambiguation of multiple mappings to the UMLS directly (105) or through WN (6). For example, the

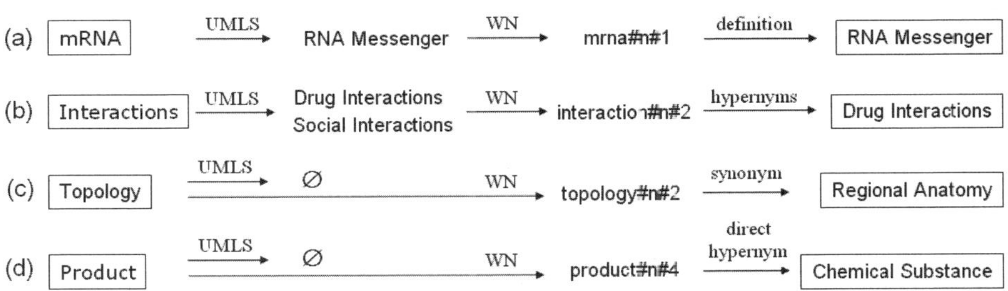

Fig. 5. Examples of cases where WN improves the direct mapping of DEs to UMLS. (a) the validation of a direct mapping; (b) the disambiguation of a multiple mapping; (c) and (d) the identification of new indirect mappings.

DE Contributor is mapped to two synsets: contributor#n#1 and contributor#n#2. The former has "Writer" and "Author" as its direct hypernyms, which both exist in UMLS. contributor#n#2 has the direct ascendant "Donor", which is also found in the UMLS. In this case, a manual review is necessary to select which of the proposed indirect mappings is correct, if any.

Instance-Level Mapping. By exploiting their values, 36 of the 548 DEs were associated with a UMLS semantic type (see Table 1). For example, the DE From, (Swiss-Prot) could not be mapped to the UMLS directly. However, its values (e.g., Rattus norvegicus, Homo sapiens) were mapped to UMLS concepts whose semantic types are descendants of Organism, indicating that the DE From represents the organism in which a protein is expressed. A new mapping between the DE From and the UMLS semantic type Organism was thus added in the global schema.

Only eleven pairs of DEs had a Jaccard similarity greater than the 0.50 threshold and were used to create additional mappings. For example, the DEs Approved Symbol (HGNC) and Gene Symbol (HGMD) have similar values (Jaccard = 0.80). Approved Symbol can thus be understood as denoting **gene** symbols (as opposed to protein symbols, for instance). Consequently, a new mapping can be identified between this DE and the UMLS concept Genes.

This example also illustrates how this method can be used for the validation of existing mappings found at the *schema* level, such as the mappings of Approved Symbol and Gene Symbol to the concept Symbols (they indeed contain **symbols**). Conversely, this method can help discover mappings wrongly identified at the *schema* level. For example, the DEs Approved Symbol and Gene Name (Entrez Gene) have a Jaccard similarity of 0.92^{14}, suggesting that one of these DEs mischaracterizes its values. After manual inspection, we determined that the values of the DE Gene Name actually correspond to gene symbols, not gene names. In this case, this method is useful for two reasons: the infelicitous mapping between the DE Gene Name and the concept Names was eliminated and a supplementary mapping was added between this DE and the concept Symbols.

3.2 Application

We created a prototype of mediator-based system, based on the elements presented above. We now present the architecture of our system, its query processing and evolution features.

Architecture and Availability. Our mediator-based system is composed of a mediator and eleven wrappers (one for each biomedical source). The mediator consists of the global schema and the set of mappings identified between concepts of the global schema and DEs. Each wrapper is composed of the local schema and the program developed for extracting DE values, which is also used for querying the corresponding source. The system is available at the following URL: http://www.med.univ-rennes1.fr/cgi-bin/mougin/These/system.pl.

---

[14] Among the 100 Web pages obtained in the two sources, each DE contains 96 non empty values and 92 are identical. Their Jaccard similarity is thus equal to 0,92.

Query Processing. The query processing includes five steps.

1. Users indicate (i) for which kind of entity they are looking (e.g., a gene name), (ii) the name(s) of the entity selected, (e.g., "hemoglobin, alpha 1" or "breast cancer, early onset"), and (iii) the type of information they want to obtain (e.g., "citation").
2. Then, the mediator identifies elements in the global schema that are relevant to the query. To this end, the mediator searches among the following terms:
   - preferred terms of concepts and semantic types;
   - synonyms of concepts in the UMLS;
   - synonyms coming from WN, if any.
3. Once these elements have been identified, query expansion is performed using the hierarchy [16]. All the descendants of the elements selected by the mediator are added to the set of concepts potentially relevant to the query. Moreover, elements whose WN hypernyms are terms of the query are also selected. Consider, for example, a biologist who is looking for comments about a given gene. No DE is associated with the term "comments". But after query expansion, the mediator selects the concept Citation (whose WN hypernyms include Comment - see Fig. 4), which, in turn, is mapped to some DEs, such as Primary Citation (PDB). Once the relevant elements have been identified in the global schema, the mediator exploits the set of mappings existing between the global schema and the DEs.
4. Then, wrappers recover the values associated with relevant DEs in each source and return them to the mediator.
5. Finally, the mediator combines the values obtained from the different sources and delivers them to users. The mediator uses the Jaccard similarity to detect similar information among DE values and eliminates redundant results.

Evolution and Scalability. Our system is designed to evolve gracefully, as the same processes used for its creation also participate in its evolution. In fact, the two major events in the evolution of a mediator-based system are the integration of a new source and changes to an existing source (Table 2).

When a new source is added, the three steps depicted in Fig. 1 have to be performed. Once general information about the new source has been collected and the program which extracts DE values from the source has been written, all the remaining tasks of the local schema acquisition are executed automatically. The mapping to the global schema is also performed automatically. A manual validation is necessary only in case of ambiguous mappings.

The update of an existing source can occur for different reasons. When the output format of results provided on the Web site changes, the program that queries this source dynamically to recover DE values has to be modified. In contrast, when the DEs of the given source have been modified, all the tasks necessary to updating the system are automatic (from DEs extraction to the modification of the global schema - see Table 2 for details).

Table 2. Summary of the steps necessary to manage the evolution of the system. Tasks performed automatically are bold-faced and for each manual task, we indicate if an interface is available to facilitate administrators' intervention. Tasks followed by a star are necessary only when a new source is added to the system.

Step	Task	Interface
	Collect of general information*	yes
Local schemas creation / modification	Creation of the program that recovers DE values*	no
	DE extraction, typing DEs, and XML schemas creation	
Mappings between local schemas and the global schema	**Direct, indirect, and through DE values**	
	Validation, if any	yes
Global schema creation / modification	**Integration of new concepts in the global schema**	

4 Discussion

Our objective was to automate as much as possible the creation and maintenance of an integration system. Toward this end, we developed methods for automatically mapping elements of sources schemas to those of the global schema. Here, we resume the contributions of our approach, discuss some of its limitations and outline how they could be addressed in future work.

4.1 Contribution of the Proposed Methods

Reuse of Existing Terminologies. The global schema of our system is based on existing terminological resources. We created it by adapting the UMLS to our needs, rather than creating a new ontology. Most existing biomedical mediator-based systems developed their own ontology, so that it suits exactly the requirements for the global schema of the integration system. For example, the developers of TAMBIS [4] created the ontology TAO [17], and designed it specifically to function as the global schema of the TAMBIS system. In contrast, we reused an independently-developed, multi-purpose terminological system, the UMLS. Reusing the UMLS was more complex, as it required us to eliminate cycles in the Metathesaurus and to determine which subset of UMLS concepts would be useful in our system.

Moreover, we enriched the global schema using WN. It actually provides complementary coverage of the DEs extracted from the eleven sources, some of which were not specific to the biomedical domain. WN thus provided additional definitions and synonyms for these concepts, and contributed to the identification of additional mappings.

Hybrid Mapping Approach. The *schema*-based approach illustrates the benefit of using an external resource to refine and complement the direct mapping strategy [18]. The use of WN indeed contributed to a substantial improvement of the results obtained by mapping DEs to the UMLS directly. Through the use of WN, the number of DEs unmapped to the global schema decreased by more than 40%. Moreover, nearly half of the unique and multiple direct mappings were validated and disambiguated, respectively.

The *instance*-based approach was useful for resolving in part the vertical integration, whose aim is to eliminate redundant data existing in biomedical sources [19]. This is a key issue that has not been addressed by existing integration systems such as TAMBIS [4], BioMediator [20], and BACIIS [21]. This approach is useful during the query process, when the mediator consolidates the results obtained from each source. The mediator simply uses the Jaccard similarity computed between pairs of DEs to detect and eliminate redundant information.

Finally, while used routinely in other domains, the combination of *schema* and *instance* approaches is original in the biomedical domain. Although underlined as necessary by [22], the exploitation of both levels had not been implemented for creating biomedical integration systems. In contrast, the hybrid approach is widespread in the artificial intelligence community, mainly for mapping schemas or ontologies [23]. More recently, it has also been exploited for integration purpose [24]. The *instance*-based approach leverages the semantics of DEs through their values. We showed that mappings obtained at the *schema* level can be valuable and that the *instance*-based approach can complement and cross-validate the traditional *schema*-based approaches.

4.2 Limitations

Query Processor. Although successful for recovering data from disparate sources automatically, our query processor could be improved. In the current implementation, the words constituting the query are mapped independently to elements of the global schema. As a consequence, some of the DEs identified as candidates to answer the query can be inappropriate. For example, a query such as "laboratory results obtained for the hepcidin gene" results (among other DEs) in the DE Mouse, Rat. This is due to the presence of "**Laboratory** Mouse" among the synonyms of the concept mapped to this DE. To address this issue, we should adapt the query process so that it considers some kind of combination of the words from the query.

The associative relations among concepts asserted in the UMLS could be added to the global schema (in complement to its hierarchical backbone) and used during the query process. In practice, neighboring concepts could used for query expansion purposes, automatically or after interactive selection by users.

The query process currently does not exploit the cross-references existing in the integrated sources. As it is done in path-based approaches [25], our system could follow the hyperlinks to recover information in other sources and provide more complete results to users. The method we developed for extracting DEs from the biomedical sources also recovers the cross-references dynamically. It would thus be possible to consider their inclusion during the query process.

Ontology Issues. Although represented in OWL DL, our global schema is not based on a formal ontology as it relies on the UMLS [26]. Other representation formalisms could have been more appropriate for describing terminological features of the UMLS. For example, SKOS (Simple Knowledge Organisation System) [27] is an emerging standard for the representation of concepts and simple structures relating concepts with associated relations (e.g., narrower than). We chose OWL DL, because it provides more expressivity and supports automatic classification [28], from which our system could benefit. In order to benefit from such services, however, we would have to enrich concepts descriptions with properties, which could be used for query reformulation. For example, in a query about proteins, the mediator would be able to eliminate concepts for entities other than proteins. Ontology-driven query reformulation would contribute to improve the accuracy of the results.

4.3 Perspectives

Enhancing Mapping Approaches. Our mapping strategy could benefit from other methods in ontology matching, surveyed in [12] and [13]. For example, the *schema*-based approach could be enhanced by the use of relations, as implemented in [29]. Indeed, the explicit relationships provided by some source vocabularies in the UMLS [30] and in WN could be exploited to refine the mappings already identified.

The results obtained through the *instance*-based approach are promising and could also be refined in several ways. The heuristics currently used for analyzing the DE values only identified a limited number of predefined types. Pattern detection could be used to identify new complex types, e.g., bibliographic references. Finally, the method used for comparing sets of values of distinct DEs could benefit from the use of learning techniques, as realized in [31].

Combination with existing systems. Some existing mediator-based systems, such as TAMBIS [4] in the biomedical domain, have developed a robust query processor. An interesting perspective could be to combine the best features of several systems. For example, creation and maintenance tasks (i.e., local schema acquisition and their mapping to the global schema) could be handled automatically by our system, while the query processing would be performed by another system, such as TAMBIS. This combination would contribute to enhance the coverage of an existing system (by feeding it with additional sources), while preserving desirable features, such as efficient query processing.

Generalization. The automatic methods proposed to create a mediator-based system should be applicable to other integration approaches. On the one hand, the method developed to acquire local schemas could be useful for the three types of integration approaches introduced in section 1. Indeed, they all require the identification of relevant information about sources, especially their schema.

On the other hand, the mapping techniques could be helpful for integration systems that include a global schema. In fact, once a global schema has been defined, it is necessary to associate its elements with those present in the local schemas. The peer-to-peer approach could particularly benefit from our work because the multiplicity of components in this type of architecture necessitates many mapping tasks among the numerous schemas [32].

In summary, we presented automated methods for creating an integration system based on the mediation approach for the biomedical domain. Existing systems show weaknesses in terms of automation of conception and evolution processes. The main contribution of this paper is to propose automated methods for acquiring sources schemas and mapping them to the global schema of the system.

Acknowledgments. This research was supported in part by the Intramural Research Program of the National Institutes of Health (NIH), National Library of Medicine (NLM).

References

1. Hernandez, T., Kambhampati, S.: Integration of Biological Sources: Current Systems and Challenges Ahead. In: Proc. ACM SIGMOD Conf., vol. 33(3), pp. 51–60 (2004)
2. Davidson, S.B., Crabtree, J., Brunk, B.P., Schug, J., Tannen, V., Overton, G.C., Stoeckert Jr., C.J.: K2/Kleisli and GUS: experiments in integrated access to genomic data sources. IBM Syst. J. 40(2), 512–531 (2001)
3. Cohen-Boulakia, S., Davidson, S.B., Froidevaux, C.: A User-Centric Framework for Accessing Biological Sources and Tools. In: Ludäscher, B., Raschid, L. (eds.) DILS 2005. LNCS (LNBI), vol. 3615, pp. 3–18. Springer, Heidelberg (2005)
4. Stevens, R., Baker, P.G., Bechhofer, S., Ng, G., Jacoby, A., Paton, N.W., Goble, C.A., Brass, A.: TAMBIS: Transparent Access to Multiple Bioinformatics Information Sources. Bioinformatics 16(2), 184–186 (2000)
5. Karp, P.D.: A Strategy for Database Interoperation. J. of Comput. Biol. 2(4), 573–583 (1995)
6. Lindberg, D.A., Humphreys, B.L., McCray, A.T.: The Unified Medical Language System. Methods Inf. Med. 32(4), 281–291 (1993)
7. McCray, A.T., Srinivasan, S., Browne, A.C.: Lexical methods for managing variation in biomedical terminologies. In: Proc Annu. Symp. Comput. Appl. Med. Care, pp. 235–239 (1994)
8. Aronson, A.R.: Effective mapping of biomedical text to the UMLS Metathesaurus: the MetaMap program. In: Proc. AMIA Symp., pp. 17–21 (2001)
9. Mougin, F., Burgun, A., Loréal, O., Le Beux, P.: Towards the automatic generation of biomedical sources schema. Medinfo. 11(2), 783–787 (2004)
10. Markowitz, V.M., Chen, I.M., Kosky, A.S., Szeto, E.: Facilities for exploring molecular biology databases on the web: a comparative study. In: Pac. Symp. Biocomput., pp. 256–267 (1997)
11. Bodenreider, O.: Circular hierarchical relationships in the UMLS: etiology, diagnosis, treatment, complications and prevention. In: Proc. AMIA Symp., pp. 57–61 (2001)
12. Rahm, E., Bernstein, P.A.: A survey of approaches to automatic schema matching. The International Journal on Very Large Data Bases 10(4), 334–350 (2001)
13. Shvaiko, P., Euzenat, J.: A survey of schema-based matching approaches. In: Spaccapietra, S. (ed.) Journal on Data Semantics IV. LNCS, vol. 3730, pp. 146–171. Springer, Heidelberg (2005)
14. Miller, G.A.: WordNet: A Lexical Database for English. ACM Communications 38(11) (1995)
15. Van Rijsbergen, C.J.: Information retrieval. Butterworth-Heinemann, Newton (1979)
16. Efthimiadis, E.N.: Query expansion. Annual review of information science and technology 31, 121–187 (1996)
17. Baker, P.G., Goble, C.A., Bechhofer, S., Paton, N.W., Stevens, R., Brass, A.: An ontology for bioinformatics applications. Bioinformatics 15(6), 510–520 (1999)
18. Zhang, S., Bodenreider, O.: Alignment of multiple ontologies of anatomy: Deriving indirect mappings from direct mappings to a reference. In: Proc. AMIA Symp., pp. 864–868 (2005)

19. Sujansky, W.: Heterogeneous database integration in biomedicine. J. Biomed. Inform. 34(4), 285–298 (2001)
20. Mork, P., Halevy, A., Tarczy-Hornoch, P.: A model for data integration systems of biomedical data applied to online genetic databases. In: Proc. AMIA Symp., pp. 473–477 (2001)
21. Ben-Miled, Z., Li, N., Liu, Y., He, Y., Lynch, E., Bukhres, O.: On the Integration of a Large Number of Life Science Web Databases. In: Rahm, E. (ed.) DILS 2004. LNCS (LNBI), vol. 2994, pp. 172–186. Springer, Heidelberg (2004)
22. Köhler, J., Philippi, S., Lange, M.: SEMEDA: ontology based semantic integration of biological databases. Bioinformatics 19(18), 2420–2427 (2003)
23. Ehrig, M., Sure, Y.: Ontology mapping - an integrated approach. In: Bussler, C.J., Davies, J., Fensel, D., Studer, R. (eds.) ESWS 2004. LNCS, vol. 3053, pp. 76–91. Springer, Heidelberg (2004)
24. Zhao, H., Ram, S.: Combining schema and instance information for integrating heterogeneous data sources. Data Knowl. Eng. 61(2), 281–303 (2007)
25. Cohen-Boulakia, S., Davidson, S.B., Froidevaux, C., Lacroix, Z., Vidal, M.E.: Path-based systems to guide scientists in the maze of biological data sources. J. Bioinform. Comput. Biol. 4(5), 1069–1095 (2006)
26. Kumar, A., Smith, B.: The Unified Medical Language System and the Gene Ontology: Some Critical Reflections. In: Günter, A., Kruse, R., Neumann, B. (eds.) KI 2003. LNCS (LNAI), vol. 2821, pp. 135–148. Springer, Heidelberg (2003)
27. Miles, A., Matthews, B., Beckett, D., Brickley, D., Wilson, M., Rogers, N.: SKOS: a language to describe simple knowledge structures for the Web. In: XTech 2005: XML, the Web and Beyond (2005)
28. Baader, F., Calvanese, D., McGuinness, D.L., Nardi, D., Patel-Schneider, P.F. (eds.): The description logic handbook: theory, implementation, and applications. Cambridge University Press, New York (2003)
29. Maedche, A., Staab, S.: Measuring Similarity between Ontologies. In: International Conference on Knowledge Engineering and Knowledge Management, pp. 251–263 (2002)
30. Schulz, S., Hahn, U.: Part-whole representation and reasoning in formal biomedical ontologies. Artificial Intelligence in Medicine 34(3), 179–200 (2005)
31. Doan, A., Madhavan, J., Domingos, P., Halevy, A.: Ontology matching: A machine learning approach. Handbook on Ontologies in Information Systems, 397–416 (2004)
32. Halevy, A.Y., Ives, Z.G., Suciu, D., Tatarinov, I.: Schema Mediation in Peer Data Management Systems. In: International Conference on Data Engineering, pp. 505–516 (2003)

VisGenome and Ensembl: Usability of Integrated Genome Maps

Joanna Jakubowska[1], Ela Hunt[2], John McClure[3], Matthew Chalmers[1],
Martin McBride[3], and Anna F. Dominiczak[3]

[1] Department of Computing Science, University of Glasgow, UK
[2] Department of Computer Science, ETH Zurich, Switzerland
[3] BHF Glasgow Cardiovascular Research Centre, University of Glasgow, UK
asia@dcs.gla.ac.uk, hunt@inf.ethz.ch,
jdmc4w@clinmed.gla.ac.uk, matthew@dcs.gla.ac.uk,
M.McBride@clinmed.ac.uk, ad7e@clinmed.gla.ac.uk

Abstract. It is not always clear how best to represent integrated data sets, and which application and database features allow a scientist to take best advantage of data coming from various information sources. To improve the use of integrated data visualisation in candidate gene finding, we carried out a user study comparing an existing general-purpose genetics visualisation and query system, Ensembl, to our new application, VisGenome. We report on experiments verifying the correctness of visual querying in VisGenome, and take advantage of software assessment techniques which are still uncommon in bioinformatics, including asking the users to perform a set of tasks, fill in a questionnaire and participate in an interview. As VisGenome offers smooth zooming and panning driven by mouse actions and a small number of search and view adjustment menus, and Ensembl offers a large amount of data in query interfaces and clickable images, we hypothesised that a simplified interface supported by smooth zooming will help the user in their work. The user study confirmed our expectations, as more users correctly completed data finding tasks in VisGenome than in Ensembl. This shows that improved interactivity and a novel comparative genome representation showing data at various levels of detail support correct data analysis in the context of cross-species QTL and candidate gene finding. Further, we found that a user study gave us new insights and showed new challenges in producing tools that support complex data analysis scenarios in the life sciences.

Keywords: visualisation of large data sets, genome maps, genome visualisation, user study, QTL, comparative and functional genomics.

1 Introduction

Data visualisation helps in the understanding of complex biological relationships, and is widely used in genomics [9,11,19,20,28], taxonomy [10], proteomics, and pathway analysis [29]. Genome data is usually served by a database system, as the amounts of data that need to be shown exceed by far the amount of

A. Bairoch, S. Cohen-Boulakia, and C. Froidevaux (Eds.): DILS 2008, LNBI 5109, pp. 77–91, 2008.

RAM available on a user machine. Significant effort goes at design time into deciding how much data to fetch from the database and how to lay it out on the screen [2,13,27]. What usually does not happen in bioinformatics is recognising the evolving needs of the visualisation user. New data types and larger volumes cause not only purely technical problems, but also perceptual ones. Adding more data 'tracks' to a visualisation, accompanied by more colours and labels, may overwhelm the user, as discussed by Catarci [5], and shown in this paper. Also, the only reliable way of anticipating and discovering user interaction problems is via a user study [7]. This paper addresses the problem of reducing the visual overload in the face of large data volumes, an issue which lies on the boundary of database and visualisation research, via a user study carried out in a controlled environment. The results of this study are being fed into further development work, and are still providing food for thought.

The motivation behind the work we report on is the need to carry out comparative analyses of QTL[1], gene and protein expression and synteny in the human, the mouse and the rat, forming part of the search for genes causing cardiovascular disease, and done in collaboration between several research groups in the UK and abroad. We first tried to find a suitable visualisation, and carried out a short study of the available browsers [18]. We discovered that the development of most browsers was not accompanied by usability studies, or such studies have not been published. We also saw that none of the viewers allowed us to see the data the way we want to view them. Expressionview [9], for example, shows QTLs and micro array probes and no other data, so it was not suitable for our work. SyntenyVista [13] shows a comparative view of two genomes but is limited with regard to other data such as micro array probes. Since the work of the British Heart Foundation Cardiovascular Research Centre at Glasgow [21] and of our collaborators requires the analysis of data of high complexity, we decided to learn from the existing packages and produce yet another genome browser. What we found missing in most browsers was the fact that it was hard to see large and small objects at the same time, and that zooming was a limiting factor. In [11] the authors recently stated explicitly that Ensembl zooming is not as flexible as maps.google.com. Since the main representational problem in our mind is zooming, this is the major technical issue we addressed, and our work examines the use of improved zooming and its contribution to the ease of traversing the genome space. We hypothesise that improved zooming will offer both usability and cognitive benefits, and aim to prove that experimentally, by comparing VisGenome and Ensembl with respect to the ease of finding of large and small objects (QTLs and micro array probes).

This paper presents the following contributions. We summarise the design and results of a user study including 15 participants which demonstrated that the users are more successful in VisGenome than in Ensembl use [16]] in the context of candidate gene analysis. Further, we discuss the findings from a user questionnaire, providing evidence that VisGenome is *perceived* to be easier to

---

[1] A quantitative trait locus (QTL) is a region of DNA that is associated with a particular phenotypic trait.

use than Ensembl. This is due to a combination of factors, including smooth zooming, provision of comparative genome views, and a simpler monochromatic display. The paper is structured as follows. Section 2 focuses on user studies in databases and bioinformatics, and Section 3 introduces VisGenome and Ensembl. Our user study design is presented in Section 4, and the results are described in Section 5. Section 6 gives a discussion, and Section 7 concludes.

2 Related Work

We first review some work spanning the areas of databases, visualisation and human computer interaction, and then summarise a number of bioinformatics user studies.

Catarci [5] was one of the first authors to convincingly argue the importance of user-centred design in the construction of user interfaces to database systems. Query construction is the focus of her work, and the design and testing process has to deliver interfaces that support efficient working and minimise user dissatisfaction and the need for assistance and maintenance. The main argument is that this can only be achieved via user-centred design, and requires the following: user involvement; a clear identification of user requirements, tasks and context; an appropriate split of functions between the user and the system; iterative design; and multidisciplinary competencies in the design team. To determine whether a system satisfies all user objectives, a formal evaluation needs to be carried out in a realistic context. As such evaluations are expensive and time-consuming, they are usually avoided, and the resulting systems are only judged in terms of correctness and functionality, and may well be suboptimal and cause user stress and additional costs to the organisation which commissioned them. One of the important points raised by Catarci is the issue of completeness and correctness of data representation. She finds that over-featured interfaces do not work well, as the complexity gets in the way of understanding the system and working out how to use it. Additionally, usability has an additional cost in terms of decrease in software production rate, and user satisfaction is never considered as an instrument to define the contract terms in software provision. As a result, also in the research context, usability issues are often ignored in favour of a narrow focus on selected information system aspects, such as performance or correctness.

A number of papers on the boundary of visualisation, e-science and database areas deal with provenance, data caching, and workflows, but address usability only in terms of user efficiency. VisTrails [4] solves the problem of visualisation from a database perspective, by managing the data and metadata of visualisation products. Workflows and provenance management are described in [22] and [6]. Here, a visualisation is used to allow the user to understand data provenance and modify existing analysis procedures (workflows). To our knowledge, no user studies have been published.

Recently, Jagadish and co-authors [15] broadened our understanding of the term usability in the context of database work. Starting from the observation that currently DBAs have to mediate between the user and the database, to hide the underlying system complexity, they draw an agenda of database usability challenges. They advocate the development of new database techniques which in their underlying design will focus on enabling direct interaction modalities for a database user. The future will be a WYSIWYG database with instantaneous-response interfaces, contextual displays, zooming and panning applying not just to maps but to all levels of database reality, including schemas, design activities, database evolution and provenance. To achieve that future, the authors propose a new presentation data model, which may be denormalised and will support direct user interaction, that is direct database creation, evolution, data manipulation, and structural changes to data. Some of the presentation modalities will include map mashups [14,8], graph representations, multidimensional database facilities and tabular metaphors for data display. In data manipulation, the user interface will take advantage of a new simple algebra that will be easy to understand and intuitive to use. The proposed research scenario includes future user studies which will guide the development of both abstract models and practical database optimisations.

We now turn our attention to bioinformatics. In this area, only a small number of user studies have been published, while many application notes and other papers published in journals *Bioinformatics* and *BMC Bioinformatics* claim that the software is 'user friendly'. For papers published in *Bioinformatics* between January 2000 and December 2007 the journal's search facility delivers 284 hits for the query 'user friendly', two for 'user study', and 53 for 'usability'. This may mean that most usability claims are not well founded. In one of the early papers mentioning the word 'user', Stevens et al. [26] presented a survey of bioinformatics tasks undertaken by biologists. They reported on new requirements which could stimulate the development of future applications, but did not conduct a user study. Wu et al. [28] reported on an electronic table that uses fisheye distortion. The table showing gene expression data was a subject of a pilot user study including five researchers completing a Questionnaire for User Interface Satisfaction (QUIS) [25]. Yang and colleagues [29] observed biologists interacting with a new software package and analysing experimental data, however, a formal study has not taken place. Graham and colleagues [10] presented an informal user study with biologists from the Royal Botanic Garden Edinburgh. The users carried out 12 tasks and used two prototypes of a visualisation tool. The authors received feedback from the participants and recognised that none of the prototypes was perfect and they should develop a new one which combined the existing two prototypes. These findings are similar to the views of the users in our study, and the feedback we obtained is reflected in our current engineering work.

3   VisGenome and Ensembl

VisGenome (VG) [17], see Figure 1 (left), shows single and comparative representations of the rat, the mouse and the human chromosomes at different levels

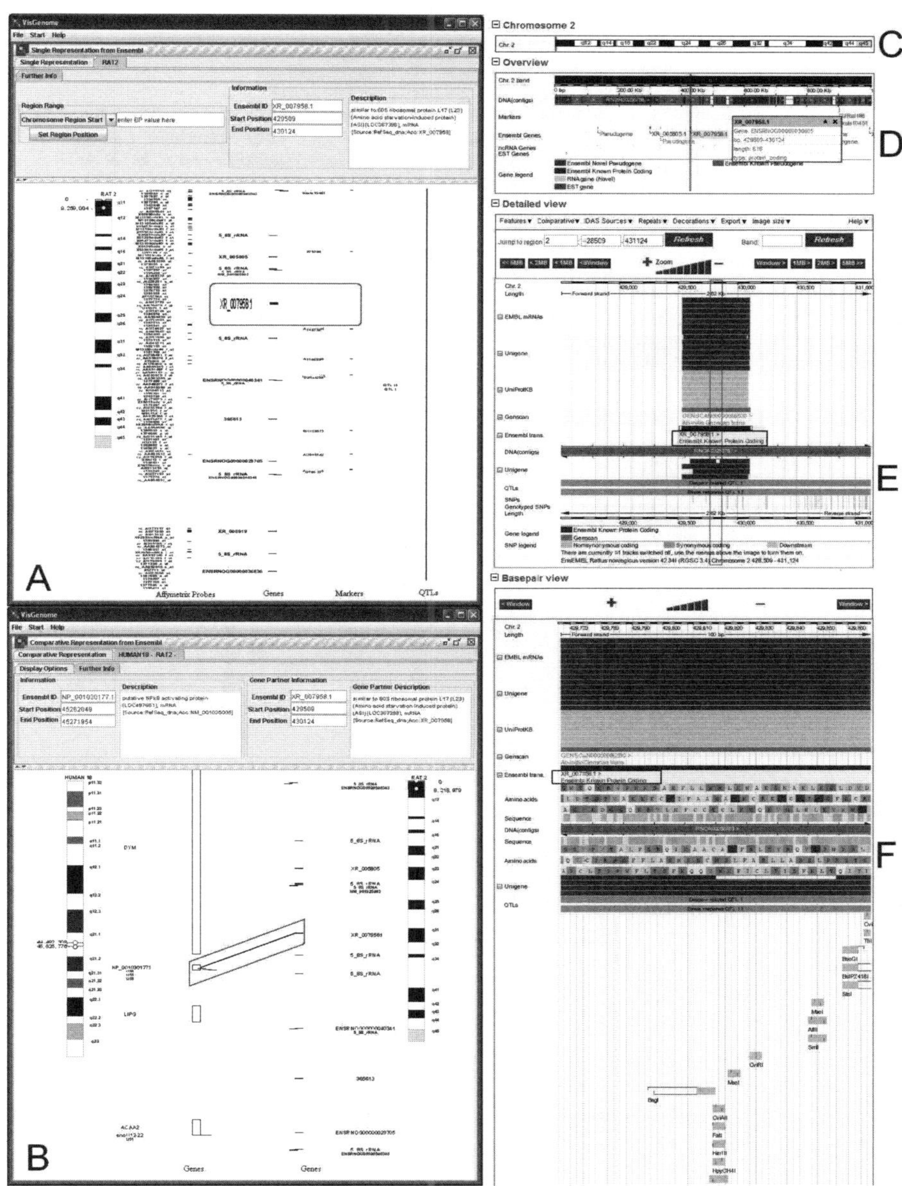

Fig. 1. Gene XR_007958.1 on rat chr. 2 in VisGenome and Ensembl, with the gene name and position in a frame superimposed on the screenshots. A: VisGenome, single chromosome view. B: VisGenome, comparative view of the rat chr. 2 and the human chr. 18. (C-F) Ensembl ContigView. (C) The entire chromosome, (D) An 'Overview' of a region of 1 Mbp, (E) The 'Detailed View' showing markers and genes, and (F) A 'Basepair View' showing protein translations.

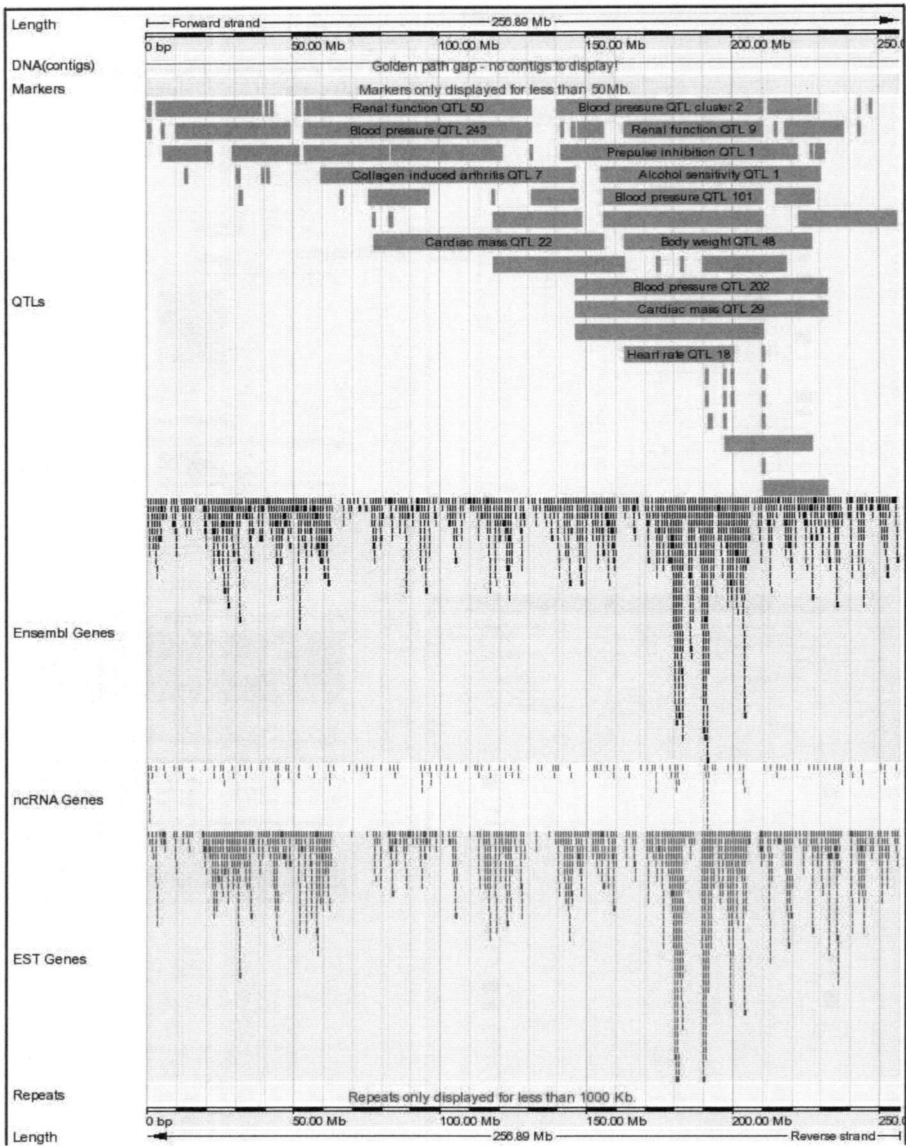

Fig. 2. Overview of rat chromosome 2 in Ensembl version 40

of detail, and integrates data from Ensembl [11] , locally produced lab results and [12]. It offers an overview of all rat, mouse and human chromosomes. After choosing a chromosome of interest, the user sees it in a new view with detailed data. The view supports interaction by mouse and keyboard, such as smooth zooming and panning [2] which is more flexible than seen in other browsers. The

Table 1. Example data sizes, Ensembl version 40, Aug 2006

Species	Chromosome	length	genes	microarray probesets	microarray probes	SNPs	QTLs
rat	2	250 Mbp	1413	1870	71,141	2740	~100

users can keep an area of interest in focus and choose the chromosome region by dragging the box enclosing the region or typing in the coordinates in an info panel. Then only the data in the selected area is displayed. The aim is to provide the context and allow the researchers to navigate the data at the same time. VG retrieves supporting web pages from Ensembl by invoking a link in a browser.

Ensembl (Ens) is probably the most popular system for mammalian genome analysis. It offers 17 different views, including ChromoView, ContigView, Gene-View, MultiContigView, SNPView, and SyntenyView. In our experiment, biological and medical researchers used ContigView, MultiContigView and SyntenyView. ContigView, see Figure 1, shows different views of a gene, from broad chromosome context to fine nucleotide detail. These views are in separate horizontal frames, one below the other, and the user has to scroll as all views do not fit on a computer screen. There is also a chromosome overview facility, CytoView, shown in Figure 2. This view does not fit on the screen either and requires scrolling. In Ensembl data items are labelled and searching on names and coordinates is possible. Zooming uses buttons (Fig. 1, panels E and F). MultiContigView is an extension of ContigView and is meant to support comparative genome analysis. It displays genome annotation for several species. In SyntenyView a clickable high-level view of chromosomes with blocks of conserved synteny is shown.

Data in Ensembl is stored in a relational database system and can be accessed via SQL or a Perl API. When the experiment was conducted, we accessed the database via JDBC and stored local experimental data and data from [12] in a local relational database. We visualised only genes, QTLs and microarray probes, and did not show SNPs or probesets, as those were not required. The requirement to show micro array probe mappings in three species increases the data size by at least a factor of 10, as each gene may have a matching micro array probe set, consisting of up to 10 probes, and each probe may have produced positive or negative results in a number of different experiments. The amount of data to be shown is significant, and is user-specific, as it may include arbitrary data sources, resulting from recent publications or experiments. Table 1 and Figure 2 give an idea of the number of items that have to be fetched from Ensembl to generate a chromosome overview, and make it clear that adding more data items and types will cause both performance and perceptual problems.

4   A User Study

The aim of the user study was to find out if new ways of visually querying the data, via mouse manipulation and zooming, are effective. Another question was whether the layout and colours we proposed supported the user in finding the data they are interested in. As our target users spend most of their time studying QTLs in the mouse, rat and human, we focus on supporting this activity, and ignore other aspects of tool use. As such work is carried out by a number of geneticists in five collaborating centres in the UK, and is poorly supported by existing tools, we wanted to see if VisGenome can facilitate it. We also wanted to gather additional feedback which would guide the development of VG. We compared Ens and VG, as Ens was the closest match to user requirements. Although the tools offer similar functions, Ens shows more data types than VG, as VG does not show sequence level data (view F in Figure 1) or gene structure (view E). VG was under our control, which allowed us to add private user data and make the study more realistic. Incorporating private data in Ensembl was not desirable, because of privacy concerns.

PARTICIPANTS. We first carried out a pilot experiment with two subjects from the Bioinformatics Research Centre (BRC) and five from the Western Infirmary (WI) in Glasgow. Finally, in the experiment we had 15 participants from the WI and the BRC. Six of them use Ens often (Ens Experts: Ex). Nine of them use different tools, such as BugView [20], UCSC GenomeBrowser [19] or AtIDB [23], or were from BRC and do not use genome browsers but know them from presentations (NonExperts: NEx). Three of the participants (Ex) previously took part in a one day Ens course.

METHODS. None of the biologists have used VG before the experiment. We gave a short presentation of VG to all subjects. Several researchers asked us to remind them first how Ens works and where to find information (three participants - NEx). We gave them a short introduction to Ens. Before the experiment, we offered the subjects the opportunity to carry out an experimental task in VG (for NEx also in Ens). We did not randomise task order and VG task came first. The order in which the tools were attempted is thus a confounding factor; although a positive effect on the performance for the second attempted tool (Ens) is the most likely consequence of this, Ens performance was not better than VG.

Prior to the study, two WI subjects had asked to see their experimental data. To that end, we created one version of VG for the majority of participants and two specific versions with private data. In those versions, micro array probes were coloured in both Single and Comparative Representations, see Figure 3. The aim was to receive more feedback from those subjects.

The experiment was divided into two parts (Ens and VG). We explained to the participants what we understand by Single and Comparative Representation and that VG offers Single and Comparative Representations, but in Ens the subjects have to decide if they would like to use MultiContigView or SyntenyView as Comparative Representation, and ContigView or any other Views as Single Representation. Some of the participants asked us if they can use BioMart [1] or RGD [24] (2 users) during the execution of Ens task. They could use all tools available from Ens pages. During the experiment the participants could give up

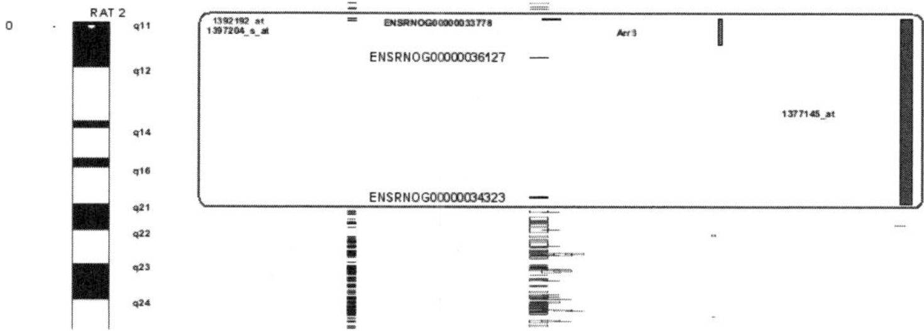

Fig. 3. VisGenome Single Representation for rat chr. 2. From left to right: chromosome overview, Affymetrix probes, genes, eQTLs and pQTLs [12], and Affymetrix probes from a user's experiment.

if they thought that it was not possible to complete the task. The majority of the subjects attempted the tasks and only one person gave up and abandoned tasks T2 and T3, see below.

SEARCH TASKS. Rather than choose our own tasks, which might have created a bias in favour of VG, we asked our biological collaborators to recommend some common search tasks. The experiment was designed to model real-life data use, and follow the pattern of an 'ecological study' under real work constraints. This precluded the use of a fully controlled experiment methodology. The users defined three tasks, as follows.

T1 Single Representation. Choose one of the rat, mouse or human chromosomes. Mark the whole chromosome and show all available data. Then choose the region between 100bp and 10,000,000bp and note the name of the first gene and the last Affymetrix probe inside the region.

T2 Comparative Representation. Choose rat chr. 18 and human chr. 5. Zoom in and out to find any homologies between genes. Then choose one of the homologies and read out the names of the homologous genes.

T3 Single Representation. Choose one of the rat chromosomes. Find the longest QTL. Then zoom on it and write down the names of the genes which are the closest to the beginning and the end of the QTL.

We captured screen usage as videos, recorded the time used for each task in minutes (STi, search time), and counted the number of mouse clicks (NoMc) for all tasks in VG and Ens. On finishing the tasks, the subjects filled in a questionnaire and participated in an interview.

5   Experimental Results

The results are quite surprising. The researchers who use Ens frequently are often unsuccessful in task execution. The experts encounter no problems in their everyday work which focuses on a chromosome fragment. However, when they

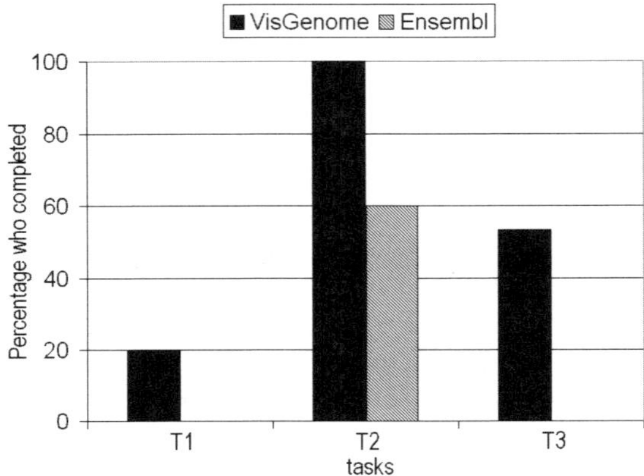

Fig. 4. Percentage of subjects (out of 15) who completed each task

examine similar data in a different part of the chromosome, they encounter problems. We also found that some of the zooming mechanisms in VG were hard to use and that the subjects prefer mouse clicking to dragging. The researchers want to see large amounts of data, but when they are looking for a particular object, they prefer to see only a small part of the data under investigation.

ACCURACY AND TASK COMPLETION. Figure 4 shows that T2, the only task involving comparative genome representation, was more successful with VG (100%) than with Ens (60%, 9 subjects). In T3 53% of attempts were successful in VG (8 subjects), while in Ens the success rate was 0. In T1 we note 20% success rate in VG and 0% in Ens. Using the two-sided sign test (where 0=both/neither successful; 1=VG success but Ens not; 1=Ens success but VG not) as an alternative to McNemar's test [3] the success rate for VG was significantly greater for both T2 (P=0.0313) and T3 (P=0.0078), but not for T1 (P=0.25). The null hypothesis for these tests was that the proportion of successes was the same for both VG and Ens, and the alternative was that they were not. Completion rates were higher in VG than in Ens for all tasks, particularly for T2 and T3. This may be due to the fact that Ens is a much richer interface, with many more options and controls and represents more data. Possibly, the subjects were not able to find out how to generate comparative genome views, or were getting lost while learning to use the system.

TIME TO FINISH. Time was measured in minutes. The biologists who completed the tasks had mean of T1=5.69' (StDev=1.39'), T2=3.58' (StDev=1.17'), and T3=5.29' (StDev=0.97') in VG and mean T2=2.83' (StDev=1.76') in Ens. As no one completed T1 and T3 in Ens, statistics were calculated only for T2. In T2 in Ens and VG 9 researchers correctly completed both tasks. As the differences in times were not normally distributed, the Wilcoxon signed rank test was used (P=0.554). We realised that Ex used both tools differently than NEx. Ex usually wanted to see more information, got interested in the data, while

NEx subjects just wanted to complete the task. Ex tried to find and show all possible answers they knew, and explore while doing the task. If there were several ways of doing the tasks in Ens they wanted to show all the solutions. In T2, for example, it was enough to show two genes in VG and Ens, and most NEx did that and finished quickly. Most Ex performed T2 and then explored MultiContigView to see more information about homologous genes, which took more time. Users behaved similarly in T3, however nobody succeeded in Ens. NEx showed Affymetrix probes in ContigView, while Ex used FeatureView and looked at the detail. There were also slight differences in server response times for Ens which might have influenced the speed of data analysis. Overall, in T2 there was little difference in task execution time between Ens and VG.

MOUSE CLICKS. Those who completed the tasks had the means of T1=53 (StDev=9.54), T2=51.07 (StDev=26.65), and T3=74.38 (StDev=13.38) NoMC in VG, and the mean for T2=23 (StDev=18.93) NoMC in Ens. Only T2 mouse clicks were analysed, due to non-completion in Ens for T1 and T3. 9 subjects completed T2 with both VG and Ens, and despite the mean number of clicks being larger in VG than in Ens, there was no significant difference in NoMC, possibly due to the small sample size. One Ex had a very large NoMC (138) for VG, and only 19 for Ens. This shows that mouse manipulation in VG needs getting used to, as panning and zooming require keeping the left/right mouse button down and moving the mouse at the same time left/right or up/down, and the left/right movement is not offered by many similar applications where clicking on zoom bars is used instead, and smooth zooming is not widely used. This is a potential problem, however, most subjects learned how to use the mouse quickly. On the other hand, Ex often clicked to see additional information and some of NEx clicked because they wanted to find the solution and they were not sure where they had to look for it. This contributed to a large NoMC in some Ex as well as NEx.

6   Discussion

6.1   User Study

In T1 we saw that the participants were looking for Affymetrix probes and couldn't find them. However, the main cause of failure in T1 was that the subjects made mistakes, e.g. typed 1 Mbp instead of 10 Mbp. In VG the subjects frequently forgot to mark the whole chromosome to show all available data or marked half of the chromosome instead of the whole. In Ens a number of users entered the coordinates and marked 'Region' instead of 'Base pair', and some did not use the overview offered by Ens but tried to mark the whole chromosome in ContigView. This usually crashed the web browser and required a restart.

T3 required showing the longest QTL. In a chromosome with many small QTLs, the subjects could not decide which QTL to choose (four subjects). We suggested that they carry out the task for any of the QTLs. The same solution was suggested where several long QTLs appeared to be of similar length. 8

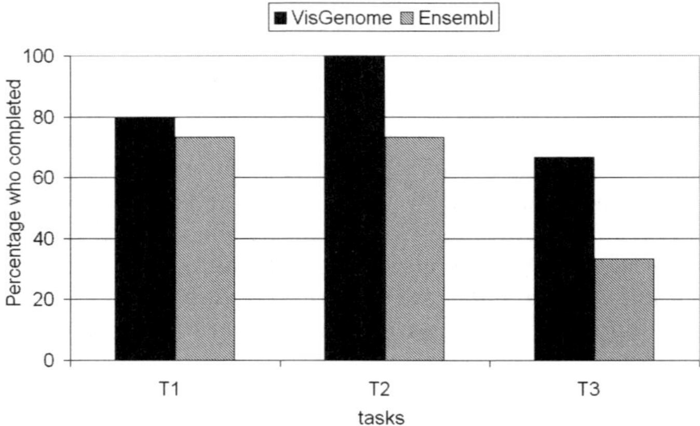

Fig. 5. Percentage of subjects (out of 15) who finished each task with errors

researchers were successful in T3 in VG. The most frequent mistake in the unsuccessful attempts in VG was choosing a complex of QTLs instead of one QTL. In Ens the subjects usually attempted to mark the entire chromosome, and only one person succeeded without crashing the browser. Some subjects tried viewing the chromosome in units of one 1 Mbp but gave up after recognising that this would take too long. One user tried to use BioMart and RGD, but this did not help. Most subjects did not realise that the view shown in Ens is not the whole chromosome but a small part of it. Several subjects chose a chromosome, clicked on it, viewed ContigView, looked down the screen to find QTLs and saw that they were all longer than the area shown in the browser, and did not know what to do to see the entire length of the QTLs.

When we analyse both correct and partially erroneous task completions, see Figure 5, we see a different view of the experiment. 11 users finished T1 and T2 in Ens and 5 users finished T3 in Ens. Similarly, for VG the completion rate improved. T1 was completed by 12 users and T3 by 10.

6.2 Lessons Learnt

Although the use of zooming helped users, and new visualisation features required some learning, we suggest that the experiment highlights another significant issue to be addressed in future development: *high error rates in data selection and query specification.* The benefits of solving this problem may well outweigh those arising from new variations on and easy learning of features such as zooming and panning. Error rates are possibly due to suboptimal menus and selection boxes, or to the fact that users find it easier to use simple interfaces with fewer options, see [5], than complex ones which offer more functionality.

We note user training is required for both VG and Ens. Although zooming and panning by mouse manipulation was classified as something very intuitive and natural, at the beginning of the VG experiment most subjects were confused and disappointed that they had to remember which button and which direction to use to zoom or pan. A possible solution to this problem would be to offer visual shortcuts to zooming, as seen in maps.google.com. While some users suggested that new visualisation techniques could be bad because biologists are not familiar with them, some said that acceptance depends on the implementation. A small number of subjects (2) suggested zooming with buttons instead of mouse manipulation and were disappointed because of the lack of scrolling.

VG supports local as well as cross-species QTL and gene expression analysis. This additional functionality offered by our application is essential to the work of our target users. In this context the use of colour will require further research, but our guess is that, based on our questionnaire, see [16], Ens offers too many colours, which is confusing to the user and makes the display hard to read. A possible extension of this work would examine the use of various layout and colouring options to arrive at solutions suitable for most users and giving the user some flexibility in layout, colour and interactivity adjustment.

Web interaction paradigms supported by AJAX (Asynchronous JavaScript and XML) are an alternative way of adding interactivity to a web-based genome map. These technologies are orthogonal to the issues of usability. We envisage that based on this study and further user studies we are planning, one could develop improved AJAX-based genome browsers which offer more interactivity and are more appropriate in the context of comparative genomics.

We confirm the findings reported in [5] about the high cost of usability experiments. The ethics application for this experiment was placed in May 2006. The user study was then refined in the summer of 2006 and conducted between August and start of December 2006. Some of the intervening time was spent on data integration tasks and some on related reading. Data analysis and writing up of the results (from screenshot recordings and questionnnaires) took about three months. This represents around 10 months of elapsed time for one PhD student, and about 1-3 hours per user. We believe the time was well spent.

7   Conclusions

We presented a user study comparing VisGenome and Ensembl in the context of comparative genome analysis. We found that in our experimental setup which targets the analysis of QTLs, synteny and gene expression, the subjects were more successful in using VG than in Ens. VG was preferable in some aspects, as it had a simpler interface, showed less data and had fewer controls. All participants liked techniques they know, such as scrolling and panning, and needed time to adapt to new solutions, such as mouse driven panning and zooming. The study shows that there is still large scope for the application of known visualisation techniques to bioinformatics data. Useful solutions, like semantic zooming offered by maps.google.com, could be very useful and should be tested in biomedical work. In particular, this study shows the great potential for usability improvement via a user study.

During the study a list of user suggestions and requests was gathered and ongoing work is addressing those, as well as exploring ways to reduce error rates in data selection and query specification. The next version of VG will be evaluated differently. We will allow the users to see their data and navigate through it. This time, instead of specified tasks, the users will use VG in a real work scenario. We will observe how the subjects interact with VG and what kind of tools and information they use. VisGenome is now usable and can be downloaded from `www.dcs.gla.ac.uk/~asia/VisGenome`. Full details of our experiment can be found in [16].

Acknowledgments. We thank all the participants for their contributions and Helen Purchase for advice in user study design. EH is an EU Marie Curie fellow, JJ is funded by the MRC, UK (grant to EH and MC), and AFD is funded by the BHF Chair and Programme Grant (BHFPG/02/128) and the Wellcome Trust Cardiovascular Functional Genomics Initiative (066780/2/012).

References

1. Ensembl BioMart, `http://www.ensembl.org/biomart/martview`
2. Bederson, B.B., Grosjean, J., Meyer, J.: Toolkit Design for Interactive Structured Graphics. IEEE Trans. Software Eng. 30(8), 535–546 (2004)
3. Bland, J.M.: An Introduction To Medical Statistics. OUP, Oxford (2000)
4. Callahan, S.P., et al.: VisTrails: visualization meets data management. In: SIGMOD Conference, pp. 745–747 (2006)
5. Catarci, T.: What Happened When Database Researchers Met Usability. Inf. Syst. 25(3), 177–212 (2000)
6. Davidson, S.B., et al.: Provenance in Scientific Workflow Systems. IEEE Data Eng. Bull. 44, 44–50 (2007)
7. Dix, A., Finlay, J., Abowd, G., Beale, R.: Human-Computer Interaction. Prentice Hall, Englewood Cliffs (2004)
8. Ennals, R., Garofalakis, M.N.: MashMaker: mashups for the masses. In: SIGMOD Conference, pp. 1116–1118 (2007)
9. Fischer, G., et al.: Expressionview: visualization of quantitative trait loci and gene-expression data in ensembl. Genome Biology 4(R477) (2003)
10. Graham, M., et al.: A Comparison of Set-Based and Graph-Based Visualisations of Overlapping Classification Hierarchies. In: Proc. AVI 2000, May 23-26, 2000, pp. 41–50 (2000)
11. Hubbard, T.J.P., et al.: Ensembl 2007. Nucleic Acids Research 35(Database-Issue), 610–617 (2007)
12. Hubner, N., et al.: Integrated transcriptional profiling and linkage analysis for identification of genes underlying disease. Nature Genetics 37, 243–253 (2005)
13. Hunt, E., Hanlon, N., Leader, D., Bryce, H., Dominiczak, A.F.: The Visual Language of Synteny. OMICS 8(4), 289–305 (2004)
14. Hunt, E., Jakubowska, J., Boesinger, C., Norrie, M.C.: Defining Mapping Mashups with BioXMash. Journal of Integrative Bioinformatics 4(3), 64 (2007); Proceedings of 4th Integrative Bioinformatics Workshop, Gent, Belgium
15. Jagadish, H.V., et al.: Making database systems usable. In: SIGMOD Conference, pp. 13–24 (2007)

16. Jakubowska, J., et al.: Usability of VisGenome and Ensembl - A User Study. Dept of Comp. Sci., University of Glasgow (2007), `http://www.dcs.gla.ac.uk/publications/PAPERS/8510/VG\&Ens_TechRep.pdf`
17. Jakubowska, J., Hunt, E., Chalmers, M., McBride, M., Dominiczak, A.F.: VisGenome: visualization of single and comparative genome representations. Bioinformatics 23(19), 2641–2642 (2007)
18. Jakubowska, J., Hunt, E., Chalmers, M.J.: Granularity of genomics data in genome visualisation. Dept of Comp. Sci., University of Glasgow (2006), `http://www.dcs.gla.ac.uk/publications/PAPERS/8212/AsiaElaMatthewCHIO6.pdf`
19. Karolchik, D., et al.: The UCSC Genome Browser Database. Nucleic Acids Research 31, 51–54 (2003)
20. Leader, D.P.: BugView: a browser for comparing genomes. Bioinformatics 20, 129–130 (2004)
21. McBride, M.W., Graham, D., Delles, C., Dominiczak, A.F.: Functional genomics in hypertension. Curr. Opin. Nephrol Hypertens 15(2), 145–151 (2006)
22. Biton, O., et al.: Zoom*UserViews: Querying Relevant Provenance in Workflow Systems. In: VLDB, pp. 1366–1369 (2007)
23. Pan, X., et al.: ATIDB: Arabidopsis thaliana insertion database. Nucleic Acids Research 31(4) (2003)
24. Pasko, D.: Overview of Rat Research Today (May 2003), `http://www.rgd.mcw.edu`
25. Slaughter, L., et al.: Assessing users' subjective satisfaction with the information system for youth services (isys). In: VA Tech Proc. 3rd Mid-Atl. Human Factors Conf., pp. 164–170 (1995)
26. Stevens, R., et al.: A classification of tasks in bioinformatics. Bioinformatics 17(2) (2001)
27. Tanin, E., et al.: Browsing large online data tables using generalized query previews. Inf. Syst. 32(3), 402–423 (2007)
28. Wu, M., et al.: A fisheye viewer for microarray-based gene expression data. BMC Bioinformatics 7(452) (2006)
29. Yang, Y., et al.: Integration of metabolic networks and gene expression in virtual reality. Bioinformatics 21(18) (2005)

An Entity Resolution Framework for Deduplicating Proteins

Lucas Lochovsky and Thodoros Topaloglou

Department of Computer Science, University of Toronto

Abstract. An important prerequisite to successfully integrating protein data is detecting duplicate records spread across different databases. In this paper, we describe a new framework for protein entity resolution, called PERF, which deduplicates protein *mentions* using a wide range of protein attributes. A *mention* refers to any recorded information about a protein, whether it is derived from a database, a high-throughput study, or literature text mining, among others. PERF can be easily extended to deduplicate protein-protein interactions (PPIs) as well. This framework translates *mentions* into instances of a reference schema to facilitate *mention* comparisons. PERF also uses "virtual attribute dependencies" to "enhance" *mentions* with additional attribute values. PERF computes a likelihood measure based upon the textual value similarity of *mention* attributes. A prototype implementation of the framework was tested, and these tests indicate that PERF can clearly separate duplicate mentions from non-duplicate *mentions*.

1   Introduction

Elucidating and cataloguing protein-protein interactions (PPIs) are important to fully understand the function and purpose of each protein in an organism's proteome. Many PPIs are now available from numerous publicly accessible databases to facilitate further research involving these interactions. Unfortunately, there are very few overlapping records between these databases [1]. Integration of this information into a single database system, however, is not straightforward, as there are many challenges to overcome in a data integration effort of this magnitude.

One particularly important data integration issue is determining which records from separate databases refer to the same actual protein [1]. This step, which is often referred to as "entity resolution" or "deduplication", is critical to ensuring that no duplicate records are present in the integrated database system. Duplicate records could be mistaken for distinct PPIs, and since these PPIs are frequently used in other analyses, quick and accurate deduplication is important to ensuring the integrity of these analyses. However, each individual PPI database usually uses its own proprietary identifier system, and therefore it is impossible to identify duplicate records by comparing identifiers. Furthermore, certain identifiers may not actually uniquely identify a single protein, but instead

A. Bairoch, S. Cohen-Boulakia, and C. Froidevaux (Eds.): DILS 2008, LNBI 5109, pp. 92–107, 2008.

refer to a class of proteins [2]. Therefore, a reliable identifier with a one-to-one correspondence to proteins is necessary in order to satisfy the goals of protein entity resolution.

In this paper, we propose a new framework for performing entity resolution on protein *mentions*. A *mention* refers to any recorded information about a protein, whether it is derived from a database, a high-throughput study, or a scientific journal, among others. PPIs can be considered pairs of protein *mentions* that interact with each other. A framework for deduplicating protein *mentions* can be easily applied to deduplicating PPIs. Given two PPIs A-B and C-D, where A-B designates an interaction between protein *mentions* A and B, if A-B and C-D refer to the same PPI, then either the pair (A,C) and the pair (B,D) are the same proteins, or (A,D) and (B,C) are the same proteins.

A reliable identifier with a one-to-one correspondence to the proteins of a given species is the Amino Acid (AA) Sequence, since the primary sequence directly determines the structure and function of each protein [3]. Therefore, if a protein *mention* provides both an AA Sequence and a "Source Organism", the one protein that this *mention* refers to can be unambiguously identified. Source Organism is required since distinct proteins in different species can share the same AA Sequence. Since the similarity of the AA Sequence and the Source Organism is generally considered to be the strongest evidence that two *mentions* refer to the same protein, existing protein deduplication systems perform deduplications solely on the basis of AA Sequence and Source Organism identity [1, 2, 4]. However, for most *mentions*, one or both of these attributes may be missing, and therefore an alternate means of deduplication is required. The new framework proposed here, the Protein Entity Resolution Framework (PERF), takes two protein *mentions* as input, attempts to deduce other attributes for these *mentions*, and makes use of these attributes to determine the likelihood that the two given *mentions* refer to the same actual protein.

PERF consists of three main components:

1. **XML Reference Schema:** The PERF framework is based on an XML schema that provides a comprehensive list of *mention* attributes derived from the schemas of various popular protein databases, including NCBI, EBI, UniProt, BIND, HPRD, MINT, MIPS, IntAct, and DIP [5, 6, 7, 8, 9, 10, 11, 12, 13, 14, 15, 16, 17]. This Framework Schema allows *mentions* to be represented in a common format to facilitate *mention* comparisons.
2. **Virtual Attribute Dependencies (VADs):** Special rules for identifying additional *mention* attributes, called "virtual attribute dependencies" (VADs), were defined for the purpose of finding as much information as possible on each *mention* to use for the actual deduplication process.
3. **Framework Deduplication Procedure:** PERF supports a computational procedure that computes the likelihood that two given protein *mentions* refer to the same actual protein based upon the attribute values available from those *mentions*.

PERF is a modular framework that currently supports the following functions:

- $resolve(m)$: This function serves as the basis for all the other functions. Given a single ambiguous *mention*, this function will resolve the protein that this *mention* refers to, if possible.
- $deduplicate(m_1, m_2)$: This function uses PERF to deduplicate two protein *mentions*. PERFs calculations should be able to identify true duplicate pairs from a set of *mention* pairs.
- $deduplicate\text{-}network(n)$: This function uses PERF to deduplicate a PPI network n, i.e. identify duplicate proteins and interactions in the network. This is essentially the application of $deduplicate(m_1, m_2)$ to each pair of *mentions* in the network to consolidate duplicate proteins and their interactions to produce a non-redundant network.
- $compare\text{-}networks(n_1, n_2)$: This function takes as input two PPI networks n_1 and n_2, both of which are internally deduplicated using $deduplicate\text{-}network(n)$. This function also finds proteins in n_1 and n_2 that are the same, and thus $compare\text{-}networks(n_1, n_2)$ can be used to determine the overlap between n_1 and n_2.

We implemented a prototype version of PERF that supports $resolve(m)$ and $deduplicate(m_1, m_2)$. We tested PERFs ability to fulfill the requirements of these functions; the test results are discussed in this paper. Although all four functions have been defined, the last two functions, $deduplicate\text{-}network(n)$ and $compare\text{-}networks(n_1, n_2)$, will be implemented for a future version of PERF.

The rest of this paper is organized as follows. Section 2 provides background information on protein and PPI database systems. Work previously done to tackle the PPI entity resolution problem is also discussed. Section 3 describes PERFs components in detail. Section 4 describes the testing of PERFs ability to fulfill the requirements of $resolve(m)$ and $deduplicate(m_1, m_2)$, and discusses the test results. Section 5 makes some concluding remarks and discusses future directions for this research.

2   Background

Many protein databases have been established to catalog all identified proteins [18]. Each of these databases relies on different sources for their records, and therefore cover very different sets of proteins. Although there is some collaboration between a few of these databases to keep each others records up-to-date, and to cross-reference corresponding records [19], most databases do not make it easy to find corresponding records in other databases. Given the exponential increase in protein data fueled by new high-throughput analyses, reliable, efficient, and automatic deduplication and integration of this data is urgently needed to properly manage this data and make sense of it.

PERF, as discussed earlier, is also applicable to the deduplication of PPIs. Many high-throughput PPI datasets have been produced in the last few years thanks to recent advances in laboratory technology [20]. These datasets are compiled from the results of high-throughput analyses. Although these analyses can process thousands of interactions in a single run, they are also prone to particularly high false positive rates (i.e. a large number of the published interactions do not actually exist) [20]. Higher confidence can be placed in interactions that are reported in several datasets, as this represents verification of these interactions in multiple, independent experiments. Therefore, reduction of false positives provides additional motivation to find duplicates and integrate high-throughput PPI datasets.

Existing protein entity resolution systems include the International Protein Index (IPI) [1], and systems like BIOZON [2] and the Agile Protein Interaction DataAnalyzer (APID) [4] for PPI deduplication. Each of these systems, however, only deduplicate proteins and PPIs on the basis of amino acid sequence similarity. PERF, however, can also make use of other available protein attributes, in addition to amino acid sequence similarity, making PERF more versatile in deduplicating protein *mentions*.

3 Protein Entity Resolution Framework (PERF)

3.1 Mentions

PERFs inputs are protein *mentions*. Typically, we refer to actual proteins with the values of their attributes, such as "Name". *Mentions* here are collections of these values drawn from a source or sources with information pertaining to a given protein. Some sources, such as database records, contain (particularly extensive) information on a given protein. Proteins may also be discussed in certain papers, either individually or within the context of a particular group of proteins. Additionally, protein information can be drawn from the data of high-throughput elucidation experiments. Each of these sources may provide different amounts and/or different types of information, but information from each of these sources is considered a *mention* for PERFs purposes.

Formally, we define a *mention* as a list of attribute-value pairs following a nested model where attributes can contain, nested within their values, "sub-attributes" or a set of values that allow lists of attributes/values to be represented within a single attribute. This model allows several aspects of a single attribute to be represented in a *mention* as well. The general form of a *mention* is described below:

$$m.name[: m.db\_name] := \{$$
$$[p_1^1 := v_1^1]\backslash;$$
$$[p_2^1 := v_2^1\backslash; v_2^2\backslash; v_2^3]\backslash;$$
$$[p_3^1 := v_3^1\backslash; [p_3^2 := v_{3-2}^1]\backslash; [p_3^3 := v_{3-3}^1\backslash; v_{3-3}^2\backslash; [p_3^4 := v_{3-4}^1]]]\backslash;$$
$$\vdots$$
$$[p_n^1 := v_n^1 \ldots]\}$$

Each *mention* is specified by a name, the name of the database it was derived from (if any), and a list of attributes. Each attribute can be associated with a single value (e.g. p_1^1), a set of values (e.g. p_2^1), or a set of sub-attributes (e.g. p_3^1). The following *mention*, which describes the CCNB1 protein from the CellMap database [21], contains examples of all three types of attributes described above.

Example 1. A complete protein *mention* using the PERF input *mention* format.

ccnb1:CellMap:={

[Name:=CCNB1]\;

[Synonyms:=Cyclin B1\;G2/mitotic specific cyclin B1\;CCNB1\;CCNB]\;

[External_Links:=[PubMed:=1387877]\;[OMIM:=123836]]\;

[Complex(s):=CDC2]\;

[Physical_Interaction(s):=CDC2\;PTCH]}

3.2   The Framework Schema

A *mention* may or may not point to a single protein entity. However, *mentions* often contain attributes that can help us retrieve additional attributes that are better suited for uniquely resolving that single protein. The Framework Schema was designed to represent these attributes in a standardized format. Therefore, given an input *mention*, we first standardize it by mapping it to the Framework Schema. Then, we expand the coverage of each *mention* with "virtual attribute dependencies" (section 3.3), and finally decide if the *mention* points to a unique protein (i.e. it is unambiguous) or a group of proteins (i.e. it is ambiguous).

The Framework Schema is a predefined XML-based schema that can accommodate many common kinds of protein information. This schema allows several instances of a *mention* to be represented in a single Framework Schema record. This is accomplished by defining the top-level element to be a "Protein_Set" that can contain multiple "Protein" objects. Initially, each Framework Schema record derived from a single *mention* contains only one "Protein". However, additional "Proteins" can be added through the use of "1-to-N VADs" described in section 3.3.

Each attribute in the Framework Schema has a distinct usefulness for the entity resolution of protein *mentions*, and therefore each attribute has been assigned a "strength". This concept resembles the selectivity of attributes in relational databases: in PERF, an attribute with strength I is a key attribute, and therefore it uniquely identifies a single protein. The less useful an attribute is for narrowing down the number of possible proteins to which a *mention* may refer, the higher its strength. The strengths of select Framework Schema attributes are provided in Table 1, along with an attribute description and the domain of accepted values for that attribute. Attribute strengths were derived from experiments with database queries to determine the cardinality of the result set produced when each attribute is used as the query attribute (databases appropriate for each attribute were used for these queries). Certain attribute combinations may be more useful for unique protein identification than the individual attributes considered in isolation; these attribute combinations are listed in Table 2. The full list of Framework Schema attributes is available in [22].

Table 1. A list of select attributes defined under the Framework Schema, along with their strengths, and the domains of the values accepted for each attribute

Attribute	Description	Strength	Domain
Name(s)	Name(s) assigned to the given protein.	III-IV	Text string corresponding to one of given proteins name(s). One tag used for each distinct name.
Keywords	Short, descriptive words assigned to given protein.	IV	Terms describing key characteristics of given protein.
Database cross-references	References to database records that describe the given protein, or some characteristic of that protein.	II	Composite value with two fields: *Name*: Name of the referenced database. *ID*: Unique identifier of referred record in named database.
Amino acid (protein) sequence	The given protein's sequence of amino acids produced by transcription and translation from the corresponding gene.	II	String of amino acid one-letter codes
Source organism	The organism from which the given protein was derived.	IV	The "[genus] [species]" designation of the source organism
Free text description	Any freeform description of the given protein.	IV	Any text
NCBI Gene ID	NCBI Gene ID identifying exact locus of gene from which given protein was transcribed.	II	A valid NCBI Gene ID

Table 2. A list of the attribute combinations that have a better strength than their separate, individual attributes, and hence have been assigned a lower number as given below

Attribute Combination	Strength
(AA Sequence, Source Organism)	I
(NT Sequence, Source Organism)	II
(NCBI Gene ID, Source Organism)	II

3.3 Virtual Attribute Dependencies (VADs)

The concept of "virtual attribute dependencies" resembles that of "functional dependencies" (FDs) in relational databases [23]. In the context of PERF, we define "virtual attribute dependencies" as rules for determining additional attribute values from an external biological database, given attribute values provided with the original *mention*. For example, if a RefSeq identifier is available in a *mention* m, then the amino acid sequence can be retrieved from the protein's RefSeq record and added to m. These newly-acquired attributes help narrow down the size of the protein classes implied by ambiguous *mentions*, and therefore the new attributes have a better strength compared to the attributes they were derived from. Formally, a virtual attribute dependency is a triple (P, Q)

$\to T$, where P refers to the set of prerequisite attributes, Q is a query or web service, and T refers to the set of resultant attributes. Given a set of values for the attributes in P, Q is evaluated to produce values for the attributes in T. Therefore, VADs define a general mechanism that is applied here to the specific problem of extending the information of protein *mentions*.

The execution of a VAD for a particular set of values for P may produce one set of values for T, or may produce many sets of values for T. Therefore, there are two types of VADs: 1-to-1 VADs and 1-to-N VADs. For the 1-to-1 VADs, the values of T are added to the original *mention* by instantiating the appropriate attributes with those values. For the 1-to-N VADs, however, each resultant value set represents one possible configuration of the original *mention*. Therefore, for each resultant value set, a new Protein object must be created in the original *mention* that extends the original Protein object with the attributes and values from that set. Thus, the *mention* is extended to cover all possible proteins that the original *mention* refers to in as much detail as possible.

VADs are designed to extend/improve an instantiation of the Framework Schema. Table 3 illustrates some example VADs. These dependencies are provided in the form $(P, Q) \to T$ described above. Starting attribute strength (Start str.) indicates the strength of the prerequisite attributes, while resultant attribute strength (Res. str.) indicates the strength of the resultant attributes. The notes column describes the rationale behind each dependency, and the last column presents examples of these dependencies with actual values. Note that this list is extensible and customizable, and can be updated to meet the deduplication needs of particular data domains.

3.4   Framework Deduplication Procedure

There are three major steps to this procedure, each of which will be discussed below.

3.4.1 Mapping Protein Mentions to the Framework Schema

Recall that the Framework Schema uses attributes names that are not the same as those of the input *mentions* but are semantically equivalent to the original *mention* attributes. A mapping procedure is therefore needed for finding the Schema attributes that correspond to a given *mention's* attributes.

Let m be a *mention* and R be the Framework Schema. Also, let $S(m)$ be the schema of m. We assume that, for each attribute a_i in $S(m)$, there is exactly one matching attribute r_j in R s.t. a_i and r_j describe the same thing. The set of these attribute pairs for each attribute a_i in $S(m)$ is called the **correct mapping**. There are two ways the Framework Deduplication Procedure can infer the correct mapping between $S(m)$ and R, depending on whether or not $S(m)$ was derived from an established database schema or not. The first option involves lexical similarity comparisons between the attributes of $S(m)$ and the attributes of R. The second option involves using a lookup table to directly translate an attribute a_i in $S(m)$ into an attribute r_j in R. This works if $S(m)$ is derived from a previously established schema that has been manually matched with the Framework Schema attributes in a one-to-one mapping. The complete description of the algorithm for inferring the correct mapping between $S(m)$ and R is available in [22].

3.4.2 Addition of Attributes to Mentions Using Virtual Attribute Dependencies

After the translation of each *mention* to a Framework Record F, the virtual attribute dependencies (VADs) in Table 3 will be used to collect additional attributes for each *mention*. Each VAD is applied sequentially, and at each step i, a Framework Record F_i is rewritten to F_{i+1}. For each VAD D_i executed on a *mention* m, the Framework Deduplication Procedure will check if all the prerequisite attributes P_i are defined in m, and if at least one of the resultant attributes T_i is not defined in m. If both of these conditions are true, then the query Q_i will be executed to produce the resultant attributes T_i to add to m. Otherwise, the next VAD will be considered, if there are any remaining VADs to consider.

3.4.3 Pairwise Matchings of Mentions

In this step of the procedure, comparisons are made between the two input mentions to determine the likelihood that they refer to the same protein. This

Table 3. A list of some of the virtual attribute dependencies (VADs) used in PERF

#	Dependency	Start str.	Res. str.	Notes	Example
1	{(Database reference), Corresponding database} → (AA Sequence, Source Organism)	II	I	All protein database records contain information on the proteins amino acid sequence, and its source organism.	{(RefSeq:= NP_660312), RefSeq} → (mmrrtlenrn ..., Homo sapiens)
2	{(NT Sequence, Source Organism), translation service} → (AA Sequence)	II	I	The nucleotide sequence can be translated into an amino acid sequence.	{(ACGAACAGGC ..., Homo sapiens), GlimmerHMM} → (malrvtrnsk ...)
3	{(AA Sequence), NCBI BLASTP} ⤳ (Source Organism) (a "⤳" means the query may or may not produce resultant attribute values, see Notes column)	IV	I	If an amino acid sequence is available, but no source organism is available, the sequence can be BLASTed against a protein database, and if a strong hit is found, and the E-value of the best hit from a different organism is lower by a threshold T than the top hit, then we can deduce the Source Organism of the uniquely identified protein referenced in the given *mention*.	{(malrvtrnsk ...), NCBI BLASTP} → (Homo sapiens)

step consists of three algorithms. They are: A) Ambiguity Determination, B) Unambiguous Deduplication, and C) Ambiguous Deduplication. Each of these will be discussed below.

A) Ambiguity Determination: Like most existing protein deduplication frameworks, we assume that AA Sequence and Source Organism are the most reliable means of identifying individual proteins [1, 2, 4]. Therefore, unambiguous *mentions* have both an AA Sequence and a Source Organism defined, and ambiguous *mentions* have one or both of these attributes undefined. If both *mentions* are unambiguous, then PERF executes an Unambiguous Deduplication (described below) that directly compares the two individual proteins, and precisely determines whether or not these proteins are the same. If one or both *mentions* are ambiguous, then there is some level of uncertainty over the protein to which one or both *mentions* refer. Under these circumstances, PERF will execute an Ambiguous Deduplication (described below) that computes a likelihood measure indicating the probability that the two *mentions* refer to the same protein.

B) Unambiguous Deduplication: In an Unambiguous Deduplication, the AA Sequence and Source Organism will be directly compared to determine if the two *mentions* describe the same protein. The sequences will be compared with the BLAST2SEQ program [24], and the organisms will be compared using the Damerau-Levenshtein (DL) string edit distance [25, 26] to determine how close they are to each other. The use of a string edit distance accommodates some tolerance for simple spelling or transcriptional errors. The results of these comparisons will be compared to cutoffs to determine if the two input *mentions* refer to the same protein. In PERFs current implementation, the BLAST2SEQ cutoff is 90% sequence identity, and the DL cutoff is 5.

C) Ambiguous Deduplication: Suppose that PERF is attempting to deduplicate two input *mentions* m_1 and m_2. Let $v(a_i, m_1)$ be the set of values of attribute a_i in *mention* m_1, and let $v(a_i, m_2)$ be the set of values of attribute a_i in *mention* m_2. For each attribute a_i in $S(m_1) \cap S(m_2)$, and any pair of *mentions* m_1 and m_2, there is a maximum number of a_i values that m_1 and m_2 can have in common. This number is the *theoretical maximum similarity score* $(M(a_i, m_1, m_2))$, and is equal to $\min\{|v(a_i, m_1)|, |v(a_i, m_2)|\}$. This is the maximum number of attribute a_i values that can match between m_1 and m_2. Attributes that are defined in one *mention* but are missing from the other are not factored into this score, since *mentions* may be derived from sources with varying attribute coverage.

The *raw similarity score* $S(a_i, m_1, m_2)$ is the actual number of a_i values that m_1 and m_2 have in common. This score is determined for each attribute a_i that has a nonzero *theoretical maximum similarity score* $M(a_i, m_1, m_2)$ on m_1 and m_2. After the calculation of the *theoretical maximum similarity score* and the *raw similarity score* between m_1 and m_2 for each a_i, the sum of the *raw similarity scores* over all attributes a_i between m_1 and m_2 is divided by the sum of the *theoretical maximum similarity scores* over all attributes a_i between m_1 and m_2 to produce a final *mention percent similarity score* $\overline{P}(m_1, m_2)$:

$$\overline{P}(m_1, m_2) = \frac{\displaystyle\sum_{a_i} S(a_i, m_1, m_2)}{\displaystyle\sum_{a_i} M(a_i, m_1, m_2)} \qquad \text{for all } a_i \text{ in } S(m_1) \cap S(m_2) \qquad (1)$$

$\overline{P}(m_1, m_2)$ will be equal to 1 if all attribute values were perfect matches, and 0 if there were no matches. In general, $0 \leq \overline{P}(m_1, m_2) \leq 1$.

So far, we have assumed that all attributes are equally important to correctly deduplicating two *mentions*. However, some might be more important than others. Therefore, we introduce an attribute weight factor. The weighted variation of the *mention percent similarity score* between m_1 and m_2 will now be discussed.

Let a be the weight factor of strength I attributes, b be the weight factor of strength II attributes, c be the weight factor of strength III attributes, and d be the weight factor of strength IV attributes. In the frameworks current form, these factors are set to the following values: $a = 1000$, $b = 100$, $c = 10$, and $d = 1$. The *weighted mention percent similarity score* between m_1 and m_2 $\overline{W}(m_1, m_2)$ is similar to the *mention percent similarity score* between m_1 and m_2 $\overline{P}(m_1, m_2)$, with the exception that the *weighted raw similarity scores* and the *weighted theoretical maximum scores* are used in the summations in the numerator and denominator, respectively. ($w(a_i)$ represents the weight factor of attribute a_i)

$$\overline{W}(m_1, m_2) = \frac{\displaystyle\sum_{a_i} w(a_i) S(a_i, m_1, m_2)}{\displaystyle\sum_{a_i} w(a_i) M(a_i, m_1, m_2)} \qquad \text{for all } a_i \text{ in } S(m_1) \cap S(m_2) \qquad (2)$$

In this algorithm, both the *mention percent similarity score* $\overline{P}(m_1, m_2)$ and the *weighted mention percent similarity score* $\overline{W}(m_1, m_2)$ are computed.

4 PERF Implementation and Evaluation

4.1 Evaluation of Mention Resolution

The International Protein Index (IPI) maintains a curated database of cross-references between a wide range of other databases, including Ensembl, RefSeq, and TAIR [1]. This index can be used to identify pairs of duplicate records across different databases. Using IPI's index, five UniProt/NCBI pairs of duplicate records were arbitrarily chosen. A set of five non-duplicate pairs was also produced by taking each of the UniProt records and randomly pairing them with NCBI records (not shown). VAD #3, which defines a rule for deriving the Source Organism of a *mention* by conducting an NCBI BLAST of the *mention's* AA Sequence (section 3.3, Table 3), was tested by removing the Source Organism from each of the UniProt *mentions*. PERF was tested on these data to determine whether or not PERF can identify the correct Source Organism, and whether or not PERF can correctly identify which pairs were actual duplicates and which were non-duplicates.

Successful invocation of VAD #3 correctly identified the Source Organism for each of the UniProt *mentions*. The results of the subsequent unambiguous deduplications demonstrate that all the actual duplicates did exhibit a sequence identity of 90% or higher, while the non-duplicates exhibited significantly worse results (data not shown). Additionally, the Source Organism DL (Damerau-Levenshtein) Distance for each pair is zero, indicating that each pair's Source Organisms were perfectly identical. Therefore, PERF correctly classified each pair in the test data, and was able to fully resolve each of the UniProt *mentions*.

4.2 Evaluation of Duplicate Resolution

The International Protein Index (IPI) was used to identify pairs of duplicate records across different databases for this evaluation. The evaluation of PERF's effectiveness at deduplicating *mention* pairs involved *mentions* derived from three of the databases for which IPI maintains cross-references. These *mention* pairs are divided into two groups representing the databases from which these *mentions* were drawn:

1. **CellMap/NCBI:** Pairs in which one *mention* was drawn from the Memorial Sloan-Kettering Cancer Center's CellMap database, and one from NCBI, and
2. **Ensembl/NCBI:** Pairs in which one mention was drawn from Ensembl, and one from NCBI

Each of these groups contains 20 arbitrarily chosen pairs of duplicate records. These *mentions* comprise the body of test cases (experiments) that PERF should correctly identify as duplicates. *Mention* pairs that do not refer to the same protein (i.e. non-duplicates) were derived by randomly pairing the NCBI *mentions* in each group to the *mentions* from the other database in the same group. (e.g. in group (i), each NCBI *mention* was randomly paired with a CellMap *mention* from the same group) Therefore, each group consists of 20 examples of *mention* pairs that refer to the same protein, and a corresponding number of examples of *mention* pairs that do not refer to the same protein. Each pair was labelled with a unique identifier indicating which group it belongs to, whether it is a duplicate or non-duplicate pair, and its unique number within that group's duplicate/non-duplicate pairs. For example, the pair II-ND-3 belongs to group II, is a non-duplicate pair, and is the third pair in the set of group II non-duplicates. All pairs from these groups were scored by the PERF Attribute Value Comparison to determine if the $\overline{W}(m_1, m_2)$ score could be used to separate the duplicate pairs from the non-duplicate pairs.

Fig. 1 presents the *mention percent similarity score* $\overline{P}(m_1, m_2)$ and *weighted mention percent similarity score* $\overline{W}(m_1, m_2)$ between m_1 and m_2 for each of the duplicate *mention* pairs and non-duplicate *mention* pairs from group I. It is clear that under the current weighting scheme, most duplicate pairs' scores are increased relative to their unweighted scores, while non-duplicate pairs scores are decreased relative to their unweighted scores. For these *mentions*, the weighting scheme slightly increased the scores of the duplicate pairs, with two exceptions. First, pair I-D-8's weighted and unweighted scores are the same. The second

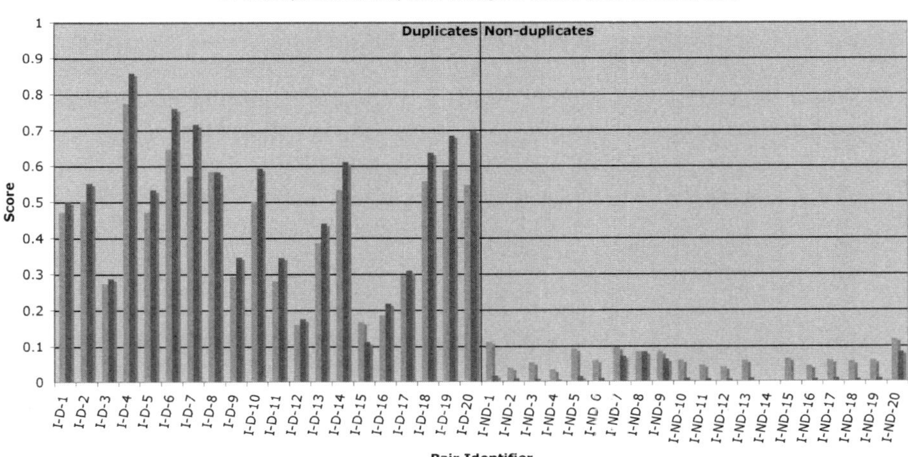

Fig. 1. Group I Results

exception is pair I-D-15, where the weighted score actually decreased relative to the unweighted score. Despite these aberrations, the scores of the duplicates are significantly higher than those of the non-duplicates. According to Fig. 1, all non-duplicate pairs' scores were drastically reduced by the weighting scheme. Therefore, overall, these weights are effective for widening the gap between actual duplicates and non-duplicates, reducing the amount of possible overlap between these two classes. Reducing this overlap is important as it reduces the number of pairs that could be mistakenly classified.

It was discovered that most of the similar attributes between duplicate *mention* pairs from this test group (i.e. between CellMap and NCBI *mentions*) were between Name attributes. Therefore, it appears that CellMap and NCBI use the same naming conventions, and Name similarity is more significant in CellMap/NCBI comparisons as a result. Consequently, the strength of Name attributes was increased when scoring these pairs.

The average $\overline{W}(m_1, m_2)$ for the duplicates was 0.497, and the average $\overline{W}(m_1, m_2)$ for the non-duplicates was 0.021. Therefore, PERF was very successful at separating true duplicates from non-duplicates. The exact score cutoff, as well as the best weighting scheme to use to separate these two classes, would be best determined by training PERF on a wider range of test data. Training could also help adjust the weighting scheme so that the weighted scores of duplicates exemplified by pairs I-D-8 and I-D-15 are increased relative to their unweighted scores.

Fig. 2 presents the $\overline{P}(m_1, m_2)$ and $\overline{W}(m_1, m_2)$ between m_1 and m_2 for each of the duplicate *mention* pairs and non-duplicate *mention* pairs from group II. Among the duplicate pairs, six pairs did not have any common attribute values, even though they actually are duplicates. (These are indicated with a Zero in

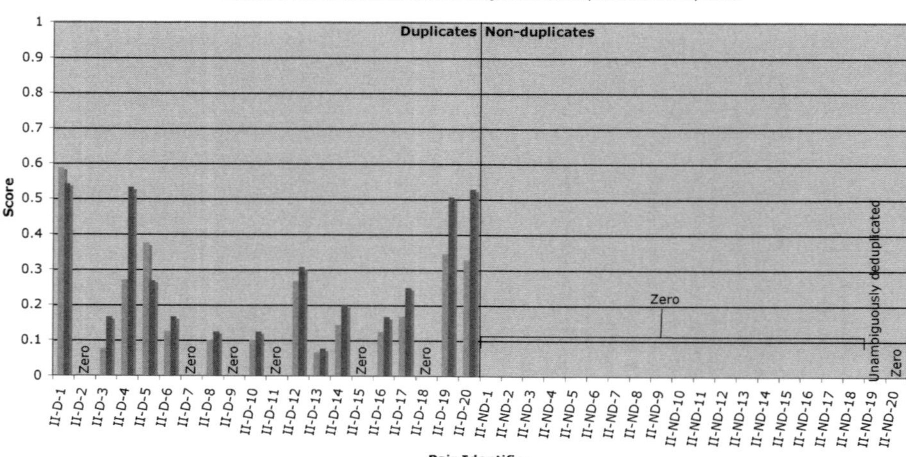

Fig. 2. Group II Results

Fig. 2) These represent duplicates that are missed, underscoring the sometimes vast differences between different databases' coverage of protein attributes. Additional attributes, possibly from the database cross-references of these *mentions*, could possibly provide attributes with similar values that PERF can identify for the purpose of establishing that these *mentions* are duplicates. PERF provides a framework where new VADs may be added to further identify new attributes. Testing with larger amounts of data in the future would help to enhance PERF capabilities in this respect.

Looking at the non-duplicate pairs, all pairs scored zero, ruling out the possibility of mistakenly classifying a non-duplicate pair as a duplicate pair. One pair, pair II-ND-19, was fully resolved by PERF, and therefore compared under the Unambiguous Deduplication Procedure described in section 3.4.3. Since the AA Sequence identity of these mentions was 24%, this pair was correctly classified as non-duplicate.

Additionally, two of the duplicate pairs from group II (pairs II-D-1 and II-D-5) have $\overline{W}(m_1, m_2)$ scores that are lower than their corresponding $\overline{P}(m_1, m_2)$ scores, much like pair I-D-15. However, as with group I, these $\overline{W}(m_1, m_2)$ scores are still adequate for distinguishing between duplicate and non-duplicate pairs. Overall, the $\overline{W}(m_1, m_2)$ scores correspond to a roughly bimodal distribution. The mean $\overline{W}(m_1, m_2)$ for the duplicates was 0.198, and the mean $\overline{W}(m_1, m_2)$ for the non-duplicates was zero, indicating that duplicates and non-duplicates are clearly separated in group II. Again, additional parameter tuning and weight training for these types of *mentions* may help produce better $\overline{W}(m_1, m_2)$ results, and help adjust the weighting scheme to increase the scores of pairs II-D-1 and II-D-5. Name attributes' strength was increased for this group as well.

The above results show that PERF is effective in deduplicating protein *mentions* by comparing a range of attributes. It is also shown that database-specific considerations are desirable for achieving a good bimodal distribution for the scores of duplicate *mention* pairs and non-duplicate *mention* pairs. *Mentions* from these databases could be used in the future as training data for PERF's attribute strengths and other database-specific parameters, allowing them to be optimized to achieve the best possible separation between duplicates and non-duplicates.

5   Conclusions and Future Work

In this paper, a new framework for deduplicating protein *mentions* was defined. Applications of this framework, PERF, to deduplicating *mention* pairs and entire networks were described. A prototype version of PERF was implemented and tested on a small set of protein *mention* pairs derived from different databases to evaluate PERF's effectiveness at fulfilling the requirements of two of the functions described earlier. These results indicate that PERF can be effective for solving the entity resolution problem for protein *mentions*. PERF forms a solid basis, grounded in techniques from database research, to address entity deduplication in biological databases.

Future plans for our work include the following developments. First, additional virtual attribute dependencies (VADs) can be produced so that there are more options available to PERF for resolving *mentions* to unique proteins. Further investigation of the Framework Schema attributes, as well as query services with which they can be used to obtain additional information, is desirable. Second, testing with larger datasets would give us more insights into increasing the effectiveness of PERF for mentions from different sources, such as published literature and high-throughput datasets. Issues specific to particular sources can also be investigated so that PERF can be better tuned for specific applications.

A third area of future development would be the creation of a better, more usable interface. Upgrading PERF to a web service will maximize its reach and enable its use by others. Fourth, there are additional steps at the end of the Framework Deduplication Procedure that could be implemented to streamline the post-deduplication user workflow. PERF could, upon completion of a deduplication, automatically consolidate two duplicate *mentions* into one, and add it to a database that serves as a repository of deduplicated *mentions*. Fifth, PERF may also have applications to the field of "data cleaning", i.e. the identification and correction of inaccurate records in a database [27]. Finally, implementations for some of the Framework Deduplication Procedure's steps could be refined to improve PERF's robustness and performance.

In conclusion, PERF forms a solid foundation for a framework for PPI deduplication. Further development of the aforementioned features, and more testing, would broaden and enhance PERF's applicability to protein and PPI deduplication problems.

References

1. Kersey, P.J., Duarte, J., Williams, A., Karavidopoulou, Y., Birney, E., Apweiler, R.: The International Protein Index: An integrated database for proteomics experiments. Proteomics 4, 1985–1988 (2004)
2. Birkland, A., Yona, G.: BIOZON: a system for unification, management and analysis of heterogeneous biological data. BMC Bioinformatics 7(70) (2006)
3. Berg, J.M., Tymoczko, J.L., Stryer, L.: Biochemistry, 5th edn. W.H. Freeman, New York (2006)
4. Prieto, C., Rivas, J.D.L.: APID: Agile Protein Interaction DataAnalyzer. Nucleic Acids Research 34, W298–W302 (2006)
5. National Center for Biotechnology Information (NCBI), http://www.ncbi.nlm.nih.gov
6. Hubbard, T., Andrews, D., Caccamo, M., Cameron, G., Chen, Y., Clamp, M., Clarke, L., Coates, G., Cox, T., Cunningham, F., Curwen, V., Cutts, T., Down, T., Durbin, R., Fernandez-Suarez, X.M., Gilbert, J., Hammond, M., Herrero, J., Hotz, H., Howe, K., Iyer, V., Jekosch, K., Kahari, A., Kasprzyk, A., Keefe, D., Keenan, S., Kokocinsci, F., London, D., Longden, I., McVicker, G., Melsopp, C., Meidl, P., Potter, S., Proctor, G., Rae, M., Rios, D., Schuster, M., Searle, S., Severin, J., Slater, G., Smedley, D., Smith, J., Spooner, W., Stabenau, A., Stalker, J., Storey, R., Trevanion, S., Ureta-Vidal, A., Vogel, J., White, S., Woodwark, C., Birney, E.: Ensembl 2005. Nucleic Acids Research 33(Database issue), D447–D453 (2005)
7. The UniProt Consortium. The Universal Protein Resource (UniProt). Nucleic Acids Research 35, 193–197 (2007)
8. Bader, G.D., Betel, D., Hogue, C.V.W.: BIND: the Biomolecular Interaction Network Database. Nucleic Acids Research 31(1), 248–250 (2003)
9. Bader, G.D., Donaldson, I., Wolting, C., Ouellette, B.F., Pawson, T., Hogue, C.W.: BINDThe Biomolecular Interaction Network Database. Nucleic Acids Research 29(1), 242–245 (2001)
10. Bader, G.D., Hogue, C.V.W.: BINDa data specification for storing and describing biomolecular interactions, molecular complexes and pathways. Bioinformatics 16(5), 465–477 (2000)
11. Peri, S., Navarro, J.D., Amanchy, R., Kristiansen, T.Z., Jonnalagadda, C.K., Surendranath, V., Niranjan, V., Muthusamy, B., Gandhi, T.K., Gronborg, M., Ibarrola, N., Deshpande, N., Shanker, K., Shivashankar, H.N., Rashmi, B.P., Ramya, M.A., Zhao, Z., Chandrika, K.N., Padma, N., Harsha, H.C., Yatish, A.J., Kavitha, M.P., Menezes, M., Choudhury, D.R., Suresh, S., Ghosh, N., Saravana, R., Chandran, S., Krishna, S., Joy, M., Anand, S.K., Madavan, V., Joseph, A., Wong, G.W., Schiemann, W.P., Constantinescu, S.N., Huang, L., Khosravi-Far, R., Steen, H., Tewari, M., Ghaffari, S., Blobe, G.C., Dang, C.V., Garcia, J.G., Pevsner, J., Jensen, O.N., Roepstorff, P., Deshpande, K.S., Chinnaiyan, A.M., Hamosh, A., Chakravarti, A., Pandey, A.: Development of human protein reference database as an initial platform for approaching systems biology in humans. Genome Research 13, 2363–2371 (2003)

12. Mishra, G., Suresh, M., Kumaran, K., Kannabiran, N., Suresh, S., Bala, P., Shivkumar, K., Anuradha, N., Reddy, R., Raghavan, T.M., Menon, S., Hanumanthu, G., Gupta, M., Upendran, S., Gupta, S., Mahesh, M., Jacob, B., Matthew, P., Chatterjee, P., Arun, K.S., Sharma, S., Chandrika, K.N., Deshpande, N., Palvankar, K., Raghavnath, R., Krishnakanth, K., Karathia, H., Rekha, B., Rashmi, N.S., Vishnupriya, G., Kumar, H.G.M., Nagini, M., Kumar, G.S.S., Jose, R., Deepthi, P., Mohan, S.S., Gandhi, T.K.B., Harsha, H.C., Deshpande, K.S., Sarker, M., Prasad, T.S.K., Pandey, A.: Human Protein Reference Database - 2006 Update. Nucleic Acids Research 34, D411–D414 (2006)
13. Chatr-aryamontri, A., Ceol, A., Palazzi, L.M., Nardelli, G., Schneider, M.V., Castagnoli, L., Cesareni, G.: MINT: the Molecular INTeraction database. Nucleic Acids Research 35(Database issue), D572–D574 (2007)
14. Munich Information Center for Protein Sequences (MIPS), http://mips.gsf.de
15. Kerrien, S., Alam-Faruque, Y., Aranda, B., Bancarz, I., Bridge, A., Derow, C., Dimmer, E., Feuermann, M., Friedrichsen, A., Huntley, R., Kohler, C., Khadake, J., Leroy, C., Liban, A., Lieftink, C., Montecchi-Palazzi, L., Orchard, S., Risse, J., Robbe, K., Roechert, B., Thorneycroft, D., Zhang, Y., Apweiler, R., Hermjakob, H.: IntAct Open Source Resource for Molecular Interaction Data. Nucleic Acids Research 35(Database issue), D561–D565 (2007)
16. Hermjakob, H., Montecchi-Palazzi, L., Lewington, C., Mudali, S., Kerrien, S., Orchard, S., Vingron, M., Roechert, B., Roepstorff, P., Valencia, A., Margalit, H., Armstrong, J., Bairoch, A., Cesareni, G., Sherman, D., Apweiler, R.: IntAct: an open source molecular interaction database. Nucleic Acids Research 32(Database issue), D452–D455 (2004)
17. Salwinski, L., Miller, C.S., Smith, A.J., Pettit, F.K., Bowie, J.U., Eisenberg, D.: The Database of Interacting Proteins: 2004 update. NAR 32(Database issue), 449–451 (2004)
18. Apweiler, R., Bairoch, A., Wu, C.H.: Protein sequence databases. Current Opinion in Chemical Biology 8, 76–80 (2004)
19. INSDC: International Nucleotide Sequence Database Collaboration, http://www.insdc.org
20. Mrowka, R., Patzak, A., Herzel, H.: Is There a Bias in Proteome Research? Genome Research 11, 1971–1973 (2001)
21. The Cancer Cell Map, http://www.cellmap.org
22. Lochovsky, L.: An Entity Resolution Framework for Deduplicating Proteins. MSc thesis. University of Toronto (2008)
23. Lee, M.L., Ling, T.W., Low, W.L.: Designing Functional Dependencies for XML. In: Jensen, C.S., Jeffery, K.G., Pokorný, J., Šaltenis, S., Bertino, E., Böhm, K., Jarke, M. (eds.) EDBT 2002. LNCS, vol. 2287, pp. 124–141. Springer, Heidelberg (2002)
24. Tatusova, T.A., Madden, T.L.: Blast 2 sequences - a new tool for comparing protein and nucleotide sequences. FEMS Microbiol Lett. 174, 247–250 (1999)
25. Damerau, F.J.: A technique for computer detection and correction of spelling errors. Communications of the ACM 7(3), 171–176 (1964)
26. Levenshtein, V.I.: Binary codes capable of correcting deletions, insertions, and reversals. Soviet Physics Doklady 10, 707 (1966)
27. Han, J., Kamber, M.: Data Mining: Concepts and Techniques. Morgan Kaufman, Burlington (2001)

Semantic Representation and Querying of caBIG Data Services

E. Patrick Shironoshita[1], Ray M. Bradley[1], Yves R. Jean-Mary[1], Thomas J. Taylor[1], Michael T. Ryan[1], and Mansur R. Kabuka[1,2]

[1] INFOTECH Soft, Inc
9200 S. Dadeland Blvd, Suite 620 Miami, Florida 33156
[2] University of Miami, Coral Gables, Florida, 33124
{patrick,rbradley,reggie,thomas,mryan,kabuka}@infotechsoft.com

Abstract. A computational grid infrastructure for biomedical research, called caGrid, is under development by the National Cancer Institute (NCI) as part of the cancer Biomedical Informatics Grid (caBIG) Initiative. In this paper we present a model that enables users to query an integrated view of caBIG data services at a conceptual semantic level. The model is based on semCDI, a formulation to generate an ontology view of caBIG semantics and pose queries against this view using the SPARQL query language complemented with Horn rules. We present here a mechanism to process these queries algebraically using our semQA query algebra extension for SPARQL, in order to create sub-expressions for each data service. We then show how resulting graphs from these sub-expressions are then merged using Horn rules.

1   Introduction

As the amount of publicly available bioinformatics and genetics data grows larger, the opportunities available to cancer biology researchers increase immensely. However, with these opportunities come additional challenges with respect to storing, retrieving, and analyzing this data [2]. It becomes essential to utilize semantic representation of the information stored in multiple data sources in order to define correspondence between entities, resolve conflicts between sources, and automate the integration process [8]. Interoperability for data representation and management can be improved as is necessary by using knowledge representation techniques that can describe the semantics of the data [13].

At the forefront of advancing the technology and implementation of semantic representations and collaborative environments is the National Cancer Institute (NCI) through its cancer Biomedical Informatics Grid (caBIG) program. caBIG is developing standards and guidelines along with data and analytical services that can be accessed and utilized through open-source software tools, all within a grid-based architecture referred to as caGrid [9].

We have developed semCDI as a model that enables users to query an integrated view of caBIG data services at a conceptual semantic level [10]. It allows researchers to utilize a single conceptual representation of the data instead of the various and

A. Bairoch, S. Cohen-Boulakia, and C. Froidevaux (Eds.): DILS 2008, LNBI 5109, pp. 108–115, 2008.

distinct domain models defined by each of the underlying data services. In this paper, we present an extension to semCDI to demonstrate the manipulation of queries into sub-expressions for specific caBIG data services using semQA, a SPARQL query algebra extension [11], and the merging of results from these sub-expressions through the application of Horn rules.

2 Background

Interoperability is addressed by caBIG using a design consisting of a syntactic layer and three semantic layers. At the syntactic layer, interface integration is handled. The semantic layers are organized as follows: first, the controlled terminology layer is maintained in the NCI Thesaurus [3], a reference terminology published by the Enterprise Vocabulary Service (EVS). It includes a list of all concepts that the caBIG semantic structure recognizes. Each of these concepts is tied to one or more common data elements (CDEs). A CDE identifies a property that can be associated with a concept; it also assigns a value restriction or value domain to that property. The third semantic layer of caBIG, the domain model layer, is data source driven, meaning that it is a collection of UML models of the caBIG compliant data services. These models are used to bind the data source metadata to caBIG's concepts and CDEs. The domain models and CDEs are contained within the cancer Data Standards Repository (caDSR). The caBIG program supports an increasing number of tools and datasets of interest to cancer research. Data services in caBIG are compatible because the elements of the underlying data sources are mapped into domain object models that are annotated to offer semantic meanings as provided in the caDSR and EVS [9]. The

Table 1. SPARQL query for genes related to the EGF pathway within caBIG

```
SELECT ?symbol ?gene_name
   ?org_name ?chr_map ?cluster_id
FROM cabio, cafe
WHERE {
  ?gene rdf:type :Gene;
   :geneSymbol ?symbol;
   :genePathway ?pathway;
   :geneOrganism ?org .
  OPTIONAL {?gene
     :geneName ?gene_name }
  OPTIONAL {?gene
     :geneLocation ?chr_map }
  OPTIONAL {?gene
     :geneClusterId ?cluster_id }
  ?org :organismName ?org_name .
  ?pathway
     :pathwayName ?pathn .
  FILTER (?pathn="h_egfPathway")
}
```

Table 2. Horn Rules for organism and gene equivalence

```
<!--rule 1-->
Organism_has_Scientific_Name
[ rdf:type ->
owl:InverseFunctionalProperty]
<!--rule 2-->
Forall ?x ?y
?x = ?y :-
And(
  ?x [ rdf:type -> :Gene ]
  ?y [ rdf:type -> :Gene ]
  Exists ?z ?w (
   And (
    ?x [ :geneOrganism -> ?z ]
    ?y [ :geneOrganism -> ?w ]
    ?z = ?w
   )
  )
  Exists ?a ?b (
   And (
    ?x [ :geneSymbol -> ?a ]
    ?y [ :geneSymbol -> ?b ]
    ?a = ?b
   )
  )
)
```

data services also require querying to be performed using CQL, an XML-based caGrid query language.

semCDI is a query formulation that defines an ontology view of caBIG semantics, where terminology concepts and domain model classes are modeled as ontology classes, associations between doman model classes are represented as object properties, attributes encoded in CDEs are modeled as datatype properties, and data objects are modeled as OWL individuals members of the corresponding domain model class. semCDI then uses the SPARQL query language [6] as the formulism to pose queries against this ontology view. Table 1 presents the SPARQL representation of a query for pathway objects representing the EGF signaling pathway, and interrelated gene and organism objects.

This merging of individuals is indicated in the semCDI query formulation using *definite Horn rules*, which define *a priori* conditional statements that are not explicitly asserted by the ontology extracted from caBIG. By design, they are defined outside of the query, to allow the use of the same rules with multiple queries independently. Two Horn rules are shown in Table 2: the first one uses the OWL inverse-functional property type to indicate that two organisms are the same if their scientific names are equal; the second rule indicates that two genes are the same if their organisms and gene symbols are equal. In these rules we use the presentation syntax derived by the Rules Interchange Format (RIF) Working Group from W3C [1].

Table 3. SPARQL algebra representation of simple query indicating source of data

```
LeftJoin(
  Join(
    Union(
      {?gene cabio:geneSymbol ?symbol},
      {?gene cafe:geneSymbol ?symbol})
    Union(
      {?gene cabio:geneName ?gene_name},
      {?gene cafe:geneName ?gene_name})),
  {?gene cabio:geneClusterId ?clusterId},
  true)
```

Table 4. SPARQL expression with zero results

```
Join(
  {?gene cabio:geneSymbol ?symbol},
  {?gene cafe:geneName ?gene_name}))
```

3 Query Processing

In order to process a query formulated against the ontology view of caBIG, it is necessary first to represent the query in terms of the data sources to be queried. This is achieved by using OWL subclasses and subproperties distinguished by namespaces. Table 3 shows a partial representation of the query in Table 1, denoted using the query algebra defined in the standard [6]. Note the use of the SPARQL Union operator where data is obtained from multiple sources. Note also that the cluster ID is only obtained from caBIO, as caFE does not contain this information.

The graph pattern expression of this rewritten query then needs to be divided into sub-expressions that obtain data from each specific data source. This is achieved by utilizing our semQA query algebra extension for SPARQL, which defines properties and equivalences between graph patterns. semQA is detailed elsewhere [11]; in summary, it substitutes Union by an idempotent-disjunction Or operator that can be distributed over both Join and LeftJoin. In particular for semCDI, algebraic manipulation is simplified by the fact that the Join of two SPARQL triple patterns from different data sources, as in Table 4, will result in zero solutions. With this, then, the query in Table 1 is transformed into the algebraic expression in Table 5.

Each sub-expression of a query so transformed is then converted into an equivalent query using CQL, the query language for caBIG; this conversion is straight-forward, and involves obtaining the domain model descriptor for each data-source specific property and class. The results from the set of CQL queries are then combined into RDF solution graphs for the original query, as described in the following section.

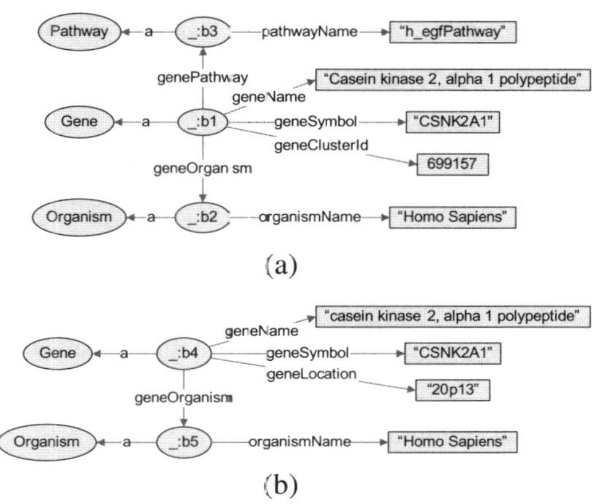

Fig. 1. RDF graph of query result from (a) caBIO and (b) caFE

Table 5. SPARQL algebra representation of query in Table 1 with sub-expressions by data source

```
or(
  Join(
    {?gene rdf:type cabio:Gene},
    {?gene cabio:geneSymbol ?symbol},
    {?gene cabio:geneName ?gene_name},
    {?gene cabio:geneClusterId ?clusterId},
    {?gene cabio:geneOrganism ?org},
    {?org cabio:organismName ?org_name},
    {?gene cabio:genePathway ?path},
    Filter((?pathn = "h_egfPathway"),
      {?path cabio:pathwayName ?pathn})),
  Join(
    {?gene rdf:type cabfe:Gene},
    {?gene cafe:geneSymbol ?symbol},
    {?gene cafe:geneName ?gene_name},
    {?gene cafe:geneLocation ?chr_map},
    {?gene cabio:geneOrganism ?org},
    {?org cabio:organismName ?org_name}}
)
```

4 Result Merging

Each CQL query results in a set of individuals that can be represented as an RDF graph. For example, the query in Table 5 results in individuals such as the ones illustrated in

Figure 1(a) for caBIO and (b) for caFE. Other similar results can be obtained from other sources that model the concept "Gene". Note that RDF blank nodes are used to denote the individuals returned.

Such solutions must be combined

Table 6. Summary of the Horn rule merge operation

```
INPUT RDF graphs g,h, query q, Horn rules r[]
m = RDF-merge(g, h)
FOR EACH variable v in query q
  FOR EACH individual x in m bound to v
    FOR EACH individual y in m bound to v
      IF x = y by some Horn rule r[i]
        z = new label not in m
        FOR EACH triple in m referring to x
          replace x by z
        FOR EACH triple in m referring to y
          replace y by z
```

according to the Horn rules associated with the query. This is achieved through the Horn-rule merge algorithm illustrated in Table 6. This algorithm takes as input two RDF graphs, g and h, a query q, and a set of Horn rules r; the two graphs are two solutions to the original query coming from different data sources. The algorithm begins with an RDF merge [7] of g and h. Next, it verifies whether two individuals within the resulting merged graph are equivalent by testing the Horn rules; if so, it merges these individuals by creating a new blank node.

The application of this algorithm to the results in Figure 1 is illustrated in Figure 2. In part (a), rule 1 from Table 2 is applied, substituting _:b2 and _:b5 as _:m1. The second part applies rule 2 to further combine _:b1 and _:b4 into _:m2. Note that in this latter combination, the ?gene_name variable could not be combined, as the lexical match between the terms is not exact due to a difference in letter case; thus, when projected into a table of variable bindings, each gene object will result in two rows, one for each name.

5 Experimental Results and Discussion

Java prototypes of the major model components have been created and interfaced with the caGrid services. The query in Table 1 was run against caBIO and caFE; other sources could not be used due to unavailability through caBIG. Table 7 shows a

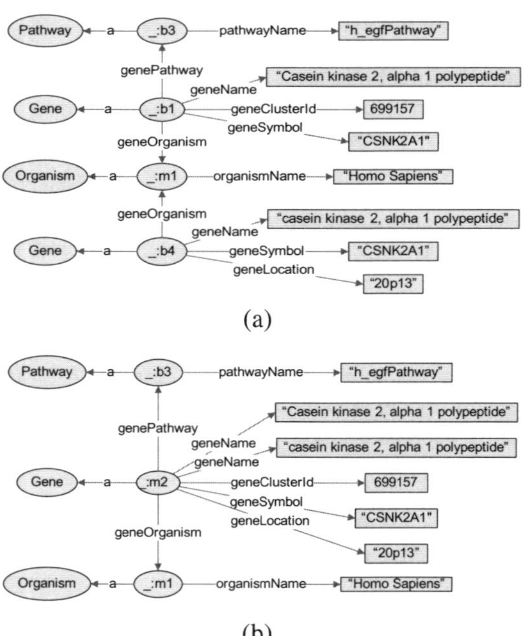

Fig. 2. RDF graph after the application of (a) rule 1 and (b) rule 2

Table 7. Partial result for the query in Table 1

symbol	?gene_name	?org_name	?chr_map	?cluster_id
FOS	V-fos FBJ murine osteosarcoma viral oncogene homolog	Homo sapiens	14q24.3	25647
FOS	v-fos FBJ murine osteosarcoma viral oncogene homolog	Homo sapiens	14q24.3	25647
CSNK2A1	Casein kinase 2, alpha 1 polypeptide	Homo sapiens	20p13	701971
CSNK2A1	casein kinase 2, alpha 1 polypeptide	Homo sapiens	20p13	701971
EGF	Epidermal growth factor (beta-urogastrone)	Homo sapiens	2q42-q43	419815
EGF	epidermal growth factor	Homo sapiens	2q42-q43	419815

listing of the first six of the 44 results obtained. We have shown a more extensive set of queries and results obtained in [10].

Several grid-specific issues were encountered. First, many of the caBIG data sources had service outages ranging from a few hours to a few days. Second, as many of these queries involve large data sets, the time it takes to formulate an ontology or receive query results ranged from two minutes to twenty minutes. Third, several of the caDSR domain models contain internal inconsistencies, for example, the field 'name' in the caFE domain model is associated with both 'organismName' and 'organismScientificName' in caDSR, causing the ambiguity to be carried over into the solution graph. As the caBIG project continues to mature and more scientists use the grid for their research, it is likely that many of these issues will be resolved by the caBIG community.

There has been recent work within the caBIG community to develop an infrastructure of data identifiers that would be used to uniquely identifying concepts on the grid. This includes work for the caGrid Identifier Services Framework, which will be capable of supplying globally unique names for caBIG domain objects. Once these additions have been implemented, our models can be extended to leverage these identifiers as a standard Horn rule applied to the result sets.

We are exploring the use of alternative caGrid query interfaces to reduce the overhead of object serialization. Also, the query model will be evaluated against more complex query patterns, and will be expanded to incorporate additional semantics available through NCI Thesaurus. Finally, a visual query interface will be designed to assist cancer researchers with query formulation.

6   Related Work

As a part of caBIG, the cancer Translational Research Informatics Platform (caTRIP) project achieves integration of data from a predefined set of grid services using an object-oriented design. caTRIP performs queries through a federated query engine by using Distributed CQL (DCQL), which is an extension to the CQL query language used by caGrid. In order to merge data from multiple sources for a single query, users must link the attributes for data elements together within the query designing portion of the user interface; the model presented here, on the other hand, uses the relationships established through Common Data Elements to define the conditions for merging data based on the semantics of caBIG.

The use of SPARQL to execute queries over distributed data nodes in a grid is proposed in [4]; unlike semCDI, this is based upon the use of relational algebra rather than algebras on SPARQL. Other mechanisms for distributing queries expressed in SPARQL and other RDF query languages include [7], [12]; these mechanisms are concerned with queries on RDF data stores. None of the systems or proposals explored in the literature utilize both SPARQL and Horn rules to enable the merging of results obtained from querying.

7 Conclusion

We have presented a model that enables users to query an integrated view of caBIG data services at a conceptual semantic level, allowing researchers to utilize a single conceptual representation of the data instead of the various and distinct domain models defined by each of the underlying data services. We have showed the application of the semCDI query formulation to generate an ontology view of caBIG semantics and define queries in SPARQL complemented with Horn rules; the use of the semQA algebra extension for SPARQL to manipulate queries and obtain sub-expressions for each data service; and the mechanism for the combination of the results from these sub-expressions according to Horn rules. We have presented examples and experimental results showing the working of the model, and we have discussed the direction of future work.

Acknowledgements

This work is supported by NIH grant 1R43CA132293. The authors also wish to acknowledge the insights given by Dr. Thomas Deisboeck at Massachusetts General Hospital, Drs. Robert Clarke and Stephen Byers at Georgetown University, and Dr. Timothy Kuzel at Northwestern University.

References

[1] Boley, H., Kifer, M.: RIF Basic Logic Dialect. W3C Working Draft 30 (October 2007), http://www.w3.org/TR/rif-bld/
[2] Collins, F.S., Green, E.D., Guttmacher, A.E., Guyer, M.S.: US National Human Genome Research Institute. A vision for the future of genomics research. Nature 422, 835–847 (2003)
[3] Hartel, F., Coronado, S., Dionne, R., Fragoso, G., Golbeck, J.: Modeling a Description Logic Vocabulary for Cancer Research. Journal of Biomedical Informatics 38, 114–129 (2005)
[4] Hayes, P. (ed.): RDF Semantics. W3C Recommendation 10 (February 2004), http://www.w3.org/TR/rdf-mt/

[5] Langegger, A., Blochl, M., Wob, W.: Sharing Data on the Grid using Ontologies and distributed SPARQL Queries. In: 18th International Workshop on Database and Expert Systems Applications, pp. 450–454. IEEE, Los Alamitos (2007)

[6] Prud'hommeaux, E., Seaborne, A.: SPARQL Query Language for RDF,
   `http://www.w3.org/TR/rdf-sparql-query/`

[7] Quilitz, B.: DARQ – Federated Queries with SPARQL (1/2008),
   `http://darq.sourceforge.net/`

[8] Rodriguez, M.A., Egenhofer, M.J.: Determining Semantic Similarity among Entity Classes from Different Ontologies. IEEE Trans. on Knowledge and Data Eng. 15, 442–456 (2003)

[9] Saltz, J., et al.: caGrid: design and implementation of the core architecture of the cancer biomedical informatics grid. Bioinformatics 22, 1910–1916 (2006)

[10] Shironoshita, E.P., Jean-Mary, Y.R., Kabuka, M.R.: semCDI: A Query Formulation for Semantic Data Integration in caBIG. JAMIA (accepted for publication)

[11] Shironoshita, E.P., Kabuka, M.R.: semQA: SPARQL Algebra and Complexity Analysis. IEE Transactions on Knowledge and Data Engineering

[12] Stuckenschmidt, H., Vdovjak, R., Broekstra, J.: Index Structures and Algorithms for Querying Distributed RDF Repositories. In: Proc. Of WWW Conf. 2004, pp. 17–22 (May 2004)

[13] Wang, X., Gorlitsky, R., Almeida, J.S.: From XML to RDF: how semantic web technologies will change the design of 'omic' standards. Nat. Biotechnol. 9, 1099–1103 (2005)

SisGen: A CORBA–Based Data Management Program for DNA Sequencing Projects

Georgios J. Pappas Jr.[1,2], Robson P. Miranda[1], Natália F. Martins[2],
Roberto C. Togawa[2], and Marcos M.C. Costa[2]

[1] Biotechnology and Genomic Sciences program, Universidade Católica de Brasília
`gpappas@bioinformatica.ucb.br`
[2] EMBRAPA Recursos Genéticos e Biotecnologia, Brazil
`mcosta@cenargen.embrapa.br`

Abstract. Biological data deluge has challenged researchers over the
last decade. Expressed sequence tag (EST) analyzes provide a rapid and
economical means to identify candidate genes, gene expression profiles
in different cell conditions, as well as functional annotation of putative
gene products. Although EST analysis tools are publicly available there
is still a lack of comprehensive data analysis and management programs.
This work presents `SisGen`, an integrated software system capable of
efficiently managing multi–user genomic projects. `SisGen` is a Java client–
server application that uses CORBA as a middleware in a multi–layer
architecture. The software integrates data management an annotation
pipeline in a rich graphical visualization environment. The architectural
design is presented and highlights the advantages in terms of portability,
interconnectivity, modularity and user interface that can be achieved
with this concept.

1 Introduction

The advent of the genomic era in the last century promoted an exponential increase in sequences in public databases, exceeding by far the capacity to perform experimental analyzes to pinpoint their roles in the cellular milieu. Concomitantly, the use of computers was perceived as pivotal to help transform sequence information into biological knowledge [1].

One type of genomic data that greatly contributed to sequence accumulation was the expressed sequence tags (ESTs; [2]). ESTs are short, single–pass sequences derived from random sequencing of cDNA library clones. Given that their generation is affordable, ESTs rapidly became a popular strategy for gene discovery in eukaryotes.

In the course of EST sequencing projects, data is continually generated by sequencing machines in the form of electropherograms, the starting material for the computational processing cascade aiming to infer biological function. Aside from numerous theoretical and algorithmic difficulties inherent in sequence annotation, a more fundamental problem of data management, processing and integration emerges. Many solutions have been developed over the years to provide

A. Bairoch, S. Cohen-Boulakia, and C. Froidevaux (Eds.): DILS 2008, LNBI 5109, pp. 116–123, 2008.

software systems dealing with the task of managing and annotating EST data, among them ESTWeb [3] and ESTExplorer [4], to cite a few. A critical evaluation several such programs was recently published [5]. A common theme is that they are web–based and coded in scripting languages like PERL or PHP. Despite progress in data organization and visualization, most of the existing systems still lack an integrated and robust approach required for EST data management.

Here we explore the concept of using enterprise–level software architectures to tackle the EST project management problem, which can be modeled as a distributed computing system. In this context, middleware technologies connect software components and provide an integration layer between heterogeneous systems. One of the earliest and most successful middleware architectures is CORBA (Common Object Request Broker Architecture), on top of which many home banking and electronic commerce systems were built.

Several groups recognized the importance of middleware technologies, such as CORBA, to enable the creation of elaborated applications integrating the many data formats and analytical tools present in the bioinformatics field [6]. Using this technology, we present a new software package, called `SisGen` that employs middleware concepts to cope with the data integration and administration problems faced in EST sequencing projects.

2   Methods

Computer Systems. All development was geared to adopt free software. The programming language and the ORB were provided by the Java Platform Standard Edition v1.5. The persistence layer was provided by the hibernate framework in conjunction with the relational database server PostgreSQL 8.2. Production servers run the Linux operating system. The clients are platform independent and distributed via Java Web Start technology. Further details can be found at `http://bioinformatics.cenargen.embrapa.br/genoma`.

Data Model. The data model was created with a project–centric vision, modeling aspects of raw sequence data being sent from different laboratories and providing detailed provenance and accounting. EST projects are hierarchically divided as having multiple cDNA libraries, each containing several plates, which in turn consist of individual reads. There is also provision for version control that permits read resubmission.

Annotation Pipeline. Starting from electropherograms, several third–party bioinformatics programs are applied in order to process raw data and provide functional annotation of the sequences. Custom–made wrappers and parsers were created to coordinate execution and integration of the ensuing results to the system. The pre–processing pipeline starts with the base calling program PHRED [7], cloning vector removal with cross_match (`http://www.phrap.org`), repeat masking with RepeatMasker (`http://www.repeatmasker.org`) and quality trimming with Lucy [8]. These steps are executed concurrently with sequence submission and provide real–time feedback to the submitter about the read/plate quality.

The next step in the pipeline is the functional annotation run on demand at server side. It starts with EST clustering using TGICL [9]. The resulting cluster consensi are subjected to several similarity searches using BLAST [10] against a series of databases defined during the project setup. Classification according to Gene Ontology (GO) and Enzyme [11] is inferred by mapping similarity search results against appropriate databases.

Sequence features that can be used as potential molecular markers for genetic studies are also annotated. Single nucleotide polymorphisms (SNPs) are predicted for each EST cluster using PolyBayes [12]. Simple sequence repeats (SSRs) are located in cluster consensi using the program `mreps` [13].

3   Results

3.1   Platform Design

The main objective was to provide an integrated software system capable of efficiently managing multi–user genomic projects, encapsulating several bioinformatics services for the analysis and manipulation of sequence data. Also, some key points were detrimental to the design process, such as (*i*) portability, (*ii*) efficiency and (*iii*) rich graphical user interface (GUI) for easy navigation.

The web based systems currently used by the vast majority of EST management systems often sacrifice design in detriment of simplicity and rapid development. Though successful most of the times, they may face shortcomings in terms of scalability, performance and flexibility. We adopted, instead, a multi–layer architecture using CORBA as middleware to service data between a java client program and the project database. Implementation was made in Java language and tried to adopt design patterns such as business object, data transfer, business delegate and session facade [14]. This promoted code reuse as well as clear separation between the data and the presentation layer. An overview of the system architecture is shown in Fig. 1, and the individual components are detailed below.

Presentation Layer. This is the piece of software used by the end user to interact with `SisGen`. Instead of a web browser, a custom made graphical user interface (GUI) was created using Java's swing library. A general overview of selected windows is shown in Fig. 2.

Departing from the common solution employing web browsers has some trade-offs though. First there is an increase in time spent designing the GUI. Also there is the versioning problem of how to distribute the client updates to maintain the compatibility with the server. This was effectively solved by using Java Web Start technology, which transparently ensures that the latest version of the application is deployed. However, the GUI programming is really an issue since the majority of the code in `SisGen` is devoted it. Notwithstanding, several benefits arise when using Java GUIs, which include better navigation and management of several windows. Also, there is a gain in flexibility since streamlined graphical components can be created, as seen in Fig. 2.

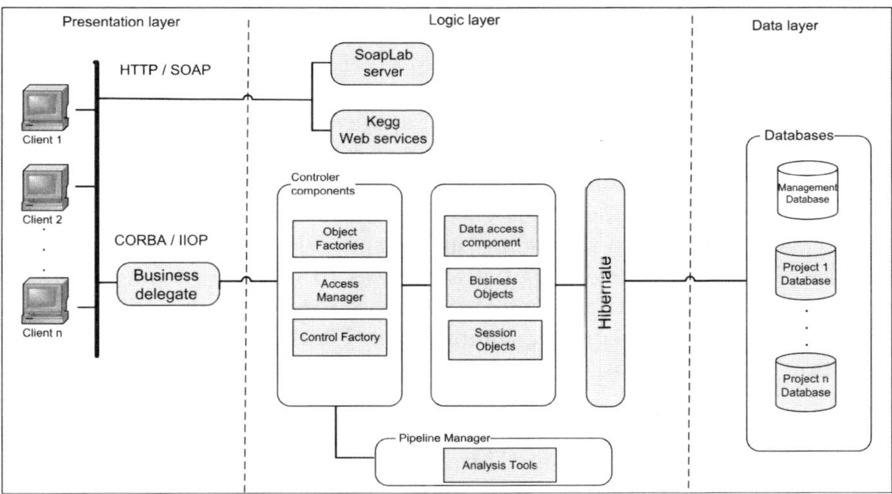

Fig. 1. Diagram of `SisGen` multi–layer architecture using CORBA as a middleware to service data between a client program and a project database. Distribution of presentation, logic and data layers elements.

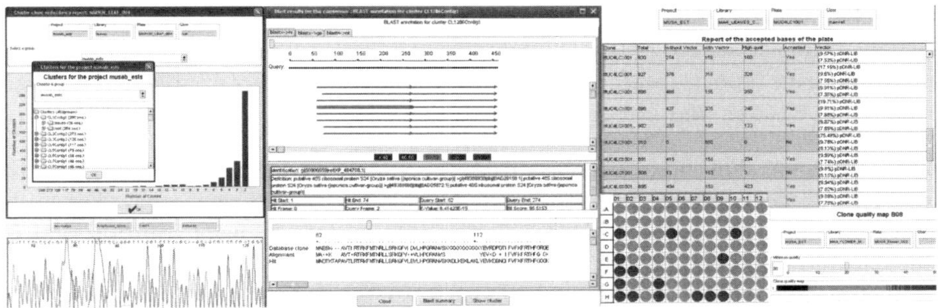

Fig. 2. Screenshot of several features available on `SisGen` showing clustering population, sequence analysis, chromatogram viewer, blast output analysis tool, plate quality visualization and clone quality map.

The client can navigate through various levels of project and sequence information, querying and gathering data from the server through the coordination of a business delegate [14], that hides client–server remote communication details (Fig. 1) reducing coupling between the presentation and logic layers.

As long as the user queries and loads data, some computation can be carried out at the client side, relieving server communication burden. In order to expand `SisGen` client capabilities, a feature was added permitting remote execution of analytical bioinformatics tools using Soaplab [15]. Soaplab exposes command–line applications as web services using SOAP (Simple Object Access Protocol)

protocol. The modular design allows `SisGen` client to seamlessly interact with different middleware technologies, aside from its core functionalities mediated by the CORBA server.

Logic Layer. Controls the main aspects of application functionality in response to client queries. It provides the unified interfaces to interact with the data layer. A object–relational mapping layer, driven by hibernate framework, hides the inner details of database operations encapsulating them in the object–oriented realm. This not only improves coding but also provides database back end independence. Additionally, one design strategy was to make provision to physical separation of the machines running the database server and the logic layer, improving security and distributing computation. Finally, the logic layer controls annotation pipeline execution, which is shielded from the end user.

Data Layer. For a specific `SisGen` project there are two main databases. One, the management database, is shared by all projects and contains project and user information details. The other database, on the same server machine, contains the sequence and annotation data itself.

3.2   System Features

Several core aspects of an EST management software are shared by `SisGen` and other web platforms, like EST data summary statistics, visualization of sequence and associated PHRED quality, project/user management or inspection of BLAST run reports, among others. Additionally, some noticeable features are peculiar to `SisGen` and are detailed next.

Data Transfer. The main use case from a sequencing facility perspective is to transfer raw electropherograms to the central bioinformatics repository. The middleware architecture adopted by `SisGen` enables a data transfer solution that is efficient and flexible. Directories, individual or compressed files containing electropherograms can be transferred in batch to the server. Real–time feedback permits the monitoring of transfer progress and individual plate quality (Fig. 3a).

Alignment Viewer. A generic sequence alignment and assembly viewer was created, capable of displaying several types of data present in an EST sequencing project. This viewer is integrated in `SisGen` but it is a completely independent and stand–alone component that can be used to visualize DNA/protein multiple alignments, BLAST results and sequence assembly files (ace format). The viewer is used to inspect the EST clusters and presents several measures of sequence conservation, like sequence logos and entropy plots (Fig. 3b).

Molecular Markers. As described in the Methods section the annotation pipeline predicts the location of polymorphic sequence sites that could be used as molecular markers. A bayesian inference procedure is used to predict the incidence of SNPs taking in consideration sequence coverage and quality [12]. An example of such SNP discovery process can be found in Fig. 3c.

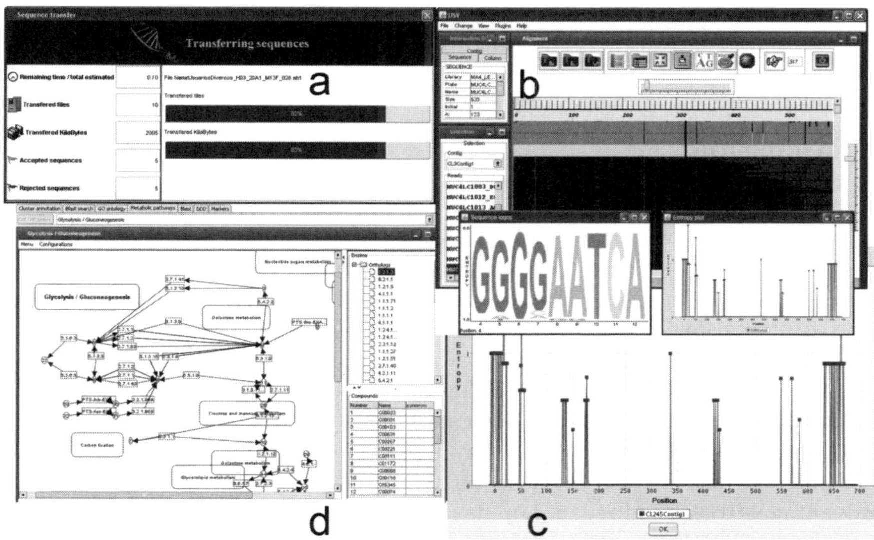

Fig. 3. Screenshots of `SisGen` user interface. a) Shows the file transfer interface; b) The Universal Sequence Viewer with cluster sequence alignment, detailing the sequence logo plug-in and the entropy plot; c) Display of Single Nucleotide Polymorphism (SNP) predictions; d) Metabolic pathway according to KEGG database.

Another type of molecular markers, the SSRs or microsatellites, are also annotated. Primer pairs flanking each microsatellite region are automatically generated. These PCR (polymerase chain reaction) pairs are suited to experimentally verify genetic diversity.

Finally, an electronic–PCR service is provided. The user provides several primer pair sequences and a search is performed to identify which sequences potentially could result in a PCR amplification product. This information can be used to assign gene annotations on markers placed on genetic maps.

Metabolic Pathway Viewer. Cross references to Enzyme database [11] are made by means of similarity searches. The sequence annotation section of `SisGen` client has an option to visualize ESTs annotated as enzymes inside their corresponding metabolic pathway(s), by performing queries to KEGG database [16]. It is possible to interrogate which ESTs map to a specific metabolic pathway and provide a visual component capable interacting with KEGG and the EST database (Fig. 3d).

3.3 Practical Applications

EST projects of varying complexities are currently being managed by `SisGen` from small to large scale. At one end the project for the plant parasite *Phytomonas serpens* contains about 2,000 ESTs from one cDNA library [17]. Conversely, the Genolyptus project [18] contains $\approx 130,000$ sequences from four eucalyptus species, obtained from more than 20 cDNA libraries, but also including genomic sequences derived from BAC (Bacterial Artificial Chromosome) ends.

4   Discussion

The inherent complexity of genomic sequencing efforts was the main motivation to create a new software for managing EST data. The multi–layer architecture centered on CORBA offers several advantages in the software engineering perspective, that sets it apart from previously reported software solutions devoted to this problem [5]. The main advantages of such design are improved modularity, efficiency and better testing and debugging. Also, the choice to create GUIs instead of using web browsers for the presentation layer, although time–consuming, pays off in terms of added capabilities of the client software to handle the heterogeneous and data–rich environment of genomics.

The core of the software, based on CORBA as the middleware, has some disadvantages tough. Albeit a popular enterprise solution last decade, several issues about CORBA complexity and maintainability were raised [19]. In our experience complexity was not an issue, since we streamlined the code to use only essential CORBA services. Still some CORBA aspects were not satisfactory, like firewall traversal and lack of ORB interoperability. The inclusion of a business delegate in our platform provides an extra level of independence from the middleware technology. In principle, porting to another middleware solution like Java RMI (Remote Method Invocation) or web services would only involve the redesign of the business delegate itself.

5   Conclusion

A new concept of EST management software is presented. It is currently in full production managing dozens of projects. In the future we envision improving data integration, by providing compatibility layers to data models such as the Generation Challenge Program standards for crop data [20] and exposing several data querying modules as BIOMOBY services [21] to enable interoperability with other bioinformatics servers.

References

1. Miller, C.J., Attwood, T.K.: Bioinformatics goes back to the future. Nature Reviews Molecular Cell Biology 4, 157–162 (2003)
2. Adams, M.D., Kelley, J.M., Gocayne, J.D., Dubnick, M., Polymeropoulos, M.H., Xiao, H., Merril, C.R., Wu, A., Olde, B., Moreno, R.F.: Complementary dna sequencing: expressed sequence tags and human genome project. Science 252, 1651–1656 (1991)
3. Paquola, A.C.M., Nishyiama, M.Y., Reis, E.M., da Silva, A.M., Verjovski-Almeida, S.: ESTWeb: bioinformatics services for EST sequencing projects. bioinformatics 19, 1587–1588 (2003)

4. Nagaraj, S.H., Deshpande, N., Gasser, R.B., Ranganathan, S.: ESTExplorer: an expressed sequence tag (EST) assembly and annotation platform. Nucleic Acids Research 35, W143–147 (2007)
5. Nagaraj, S.H., Gasser, R.B., Ranganathan, S.: A hitchhiker's guide to expressed sequence tag (EST) analysis. Briefings in Bioinformatics 8, 6–21 (2007)
6. Stevens, R., Miller, C.: Wrapping and interoperating bioinformatics resources using CORBA. Briefings in Bioinformatics 1, 9–21 (2000)
7. Ewing, B., Hillier, L., Wendl, M.C., Green, P.: Base–calling of automated sequencer traces using phred. i. accuracy assessment. Genome Research 8, 175–185 (1998)
8. Chou, H.H., Holmes, M.H.: DNA sequence quality trimming and vector removal. Bioinformatics 17, 1093–1104 (2001)
9. Pertea, G., Huang, X., Liang, F., Antonescu, V., Sultana, R., Karamycheva, S., Lee, Y., White, J., Cheung, F., Parvizi, B., Tsai, J , Quackenbush, J.: TIGR gene indices clustering tools (TGICL): a software system for fast clustering of large EST datasets. Bioinformatics 19, 651–652 (2003)
10. Altschul, S.F., Madden, T.L., Schaffer, A.A., Zhang, J., Zhang, Z., Miller, W., Lipman, D.J.: Gapped BLAST and PSI-BLAST: a new generation of protein database search programs. Nucleic Acids Research 25, 3389–3402 (1997)
11. Bairoch, A.: The ENZYME database in 2000. Nucleic Acids Research 28, 304–305 (2000)
12. Marth, G.T., Korf, I., Yandell, M.D., Yeh, R.T., Gu, Z., Zakeri, H., Stitziel, N.O., Hillier, L., Kwok, P.Y., Gish, W.R.: A general approach to single–nucleotide polymorphism discovery. Nature Genetics 23, 452–456 (1999)
13. Kolpakov, R., Bana, G., Kucherov, G.: mreps: Efficient and flexible detection of tandem repeats in DNA. Nucleic Acids Research 31, 3672–3678 (2003)
14. Marinescu, F.: Ejb Design Patterns: Advanced Patterns, Processes, and Idioms. John Wiley & Sons, Inc., New York (2002)
15. Senger, M., Rice, P., Oinn, T.: Soaplab - a unified sesame door to analysis tools, 509–513 (2003)
16. Kanehisa, M., Goto, S.: KEGG: kyoto encyclopedia of genes and genomes. Nucleic Acids Research 28, 27–30 (2000)
17. Pappas, G.J., Benabdellah, K., Zingales, B., González, A.: Expressed sequence tags from the plant trypanosomatid Phytomonas serpens. Molecular and Biochemical Parasitology 142, 149–157 (2005)
18. Grattapaglia, D.: Integrating genomics into eucalyptus breeding. Genetics and Molecular Research 3, 369–379 (2004)
19. Henning, M.: The rise and fall of CORBA. Queue 4, 28–34 (2006)
20. Bruskiewich, R., Davenport, G., Hazekamp, T., Metz, T., Ruiz, M., Simon, R., Takeya, M., Lee, J., Senger, M., McLaren, G., Hintum, T.V.: Generation challenge programme (GCP): standards for crop data. Omics 10, 215–219 (2006)
21. Wilkinson, M.D., Links, M.: BioMOBY: an open source biological web services proposal. Briefings in Bioinformatics 3, 331–341 (2002)

Bgee: Integrating and Comparing Heterogeneous Transcriptome Data Among Species

Frederic Bastian[1,2,*], Gilles Parmentier[1,2,*], Julien Roux[1,2], Sebastien Moretti[1,2], Vincent Laudet[3], and Marc Robinson-Rechavi[1,2]

[1] Department of Ecology and Evolution, University of Lausanne, quartier UNIL-Sorge, 1015 Lausanne, Switzerland
[2] Swiss institute of bioinformatics, Lausanne, Switzerland
[3] Université de Lyon, Institut de Génomique Fonctionnelle de Lyon, ENS Lyon, Université Lyon 1, CNRS, INRA, Institut Fédératif 128 Biosciences Gerland Lyon Sud, France
Marc.Robinson-Rechavi@unil.ch

Abstract. Gene expression patterns are a key feature in understanding gene function, notably in development. Comparing gene expression patterns between animals is a major step in the study of gene function as well as of animal evolution. It also provides a link between genes and phenotypes. Thus we have developed Bgee, a database designed to compare expression patterns between animals, by implementing ontologies describing anatomies and developmental stages of species, and then designing homology relationships between anatomies and comparison criteria between developmental stages. To define homology relationships between anatomical features we have developed the software Homolonto, which uses a modified ontology alignment approach to propose homology relationships between ontologies. Bgee then uses these aligned ontologies, onto which heterogeneous expression data types are mapped. These already include microarrays and ESTs. Bgee is available at http://bgee.unil.ch/

Keywords: gene expression pattern, homology, ontology, data integration.

1 Introduction

Gene expression patterns (when and where a gene is expressed) are a key feature that underlies the development of organisms and phenotypes of individuals. They are an important aspect of the study of gene function. Moreover, the study of the evolution of developmental processes, often called "evo-devo", has shown that the primary source of change in the evolution of phenotypes is changes in gene expression [1] rather than sequence.

Comparing gene expression patterns between animals is thus a major step in the study of gene function as well as of animal evolution, and also provides a link between genes and phenotypes.

---

* Co first authors.

A. Bairoch, S. Cohen-Boulakia, and C. Froidevaux (Eds.): DILS 2008, LNBI 5109, pp. 124–131, 2008.

In biological research, results obtained in different organisms are routinely compared. A comparative approach may be chosen for practical reasons because the organism of interest (humans, farm animals) may be less amenable to experimentation than more or less distant model species (as mouse, rat, zebrafish, or fruit fly).

Another reason is that components of gene expression may vary for no obvious reason [2]; this introduces the problem of distinguishing this signal from the noise caused both by random evolution and the inaccurate data measurements. Comparative study of gene expression in several species may contribute to this distinction. For example, comparing multiple samples from humans and rodents gave sufficient statistical evidence for a functionally relevant component of gene expression [3], and allowed for significant improvement in tumour characterisation [4].

Transcriptome data have also been compared among species to gain direct insight into evolutionary processes. For instance, yeast microarray data provided evidence for divergence of expression after genome duplication [5], and further studies have succeeded in extracting some evidence for the evolution of new gene functions after genome duplication in yeast and human lineages [6, 7]. A comparative approach would allow to understand the mechanisms and the consequences of gene expression evolution.

We have developed Bgee (a dataBase for Gene Expression Evolution) to address these questions. Bgee must answer the following requirements, to enable large scale gene expression pattern comparison:

- Precise description of the anatomy and developmental stages of each species, stored in a computer-understandable way.
- Integration of expression data in order to know in which anatomical features (spatial mapping) and which developmental stages (temporal mapping) genes are expressed.
- Comparison criteria between anatomies, developmental stages, and genes.

To unambiguously describe anatomy and development of a species in a computer-understandable way, ontologies are required: they describe a domain of knowledge, by using well-defined concepts and designing relationships amongst them. Several databases provide species-specific ontologies that describe anatomical features for a species, such as ZFIN [8] for the zebrafish. But as far as we know, no database provides relationships between these ontologies to allow comparisons.

The appropriate criterion to make comparisons in an evolutionary context is homology: we need to compare features that derive from the same ancestral element. We have thus designed homology relationships between anatomies of different species. This is a difficult task, and Bgee implements computational methods to achieve it (section 2). Then, we need homology relationships between genes. This point has already been abundantly treated in bioinformatics, and will not be discussed in detail in this paper. Finally, we need relationships between developmental stages. As these stages are artificial features that help to describe the continuous process of development, homology cannot be defined in a rigorous manner. We have rather designed a mapping of "equivalent" developmental stages between species (section 3).

To describe gene expression patterns, Bgee requires large amounts of data. To this end, heterogeneous data types are used (ESTs, microarrays, and soon *in situ* hybridizations). The common information to gather is whether an experiment has determined that a gene is expressed or not, and with which confidence. We have applied different statistical tests for each data type to obtain this information (section 4).

Thanks to the successful implementation of all these requirements (anatomical and developmental ontologies, comparison relationships between ontologies and genes, integration of heterogeneous expression data), Bgee allows the easy retrieval of gene expression data for different species, as well as the automated comparison of gene expression patterns.

2 Designing Homology Relationships between Anatomical Ontologies by an Ontology Alignment Approach

To study the evolution of gene expression patterns, comparisons have to be done between organs that evolved from a common ancestral structure. Thus designing relationships between anatomical ontologies consists in finding correspondences (homology relationships) between the concepts (organs) of these ontologies. This problem is a special case of "schema matching", or "ontology alignment".

Ontology alignment ([9] for a review) is the process of determining correspondences between ontology concepts. Usually, this technique is used to find the common concepts present in two ontologies. In the case of anatomical ontologies, the concepts to align are not strictly common, but rather, related: a homology relationship is not an equivalence relationship. For this reason, ontology alignment approaches developed for other applications cannot be applied as is: these methods would be misled by the existence of elements of same names and related to the same concept, but not homologous (eye of insects and of vertebrates for instance), or reciprocally, homologous elements with different names (pectoral fin and upper limb for instance). This is why we apply modified ontology alignment techniques in order to find putative homologies between two species anatomies. An expert has to manually validate the putative homologs. This method is implemented by Homolonto, a software that we have developed in Java. Homolonto will be presented in detail elsewhere; we present here the outline of its algorithm.

Our process is a supervised one: at each step, some homology relationships are proposed to the expert, who may validate them or not. Computations are made based on these decisions, and new propositions are made to the expert.

The algorithm starts with a list of pairs, which have identical names. This is based on the assumption that two structures that have the same name are likely homologous. For example, "optic cup" of the ZFIN ontology (zebrafish) and "optic cup" of the EHDA ontology (human) will be paired, but "optic cup" of ZFIN will not be initially paired with "optic nerve" of EHDA. The score of similarity between terms is up weighted by the proportion of common words, and down weighted by the frequency of these words (frequent words are less informative, e.g. "endoderm"). Moreover, scores are propagated between pairs which are neighbors in both ontologies. For example, the score of the "optic cup" pair is added to the score of the "eye" pair, as "optic cup" is part of "eye". In the same way, the score of the "eye" pair is added to the "optic cup" one.

Each pair is proposed to the expert, in descending order of scores. The expert may validate or invalidate the hypothesis of homology, or delay decision. The expert may choose to evaluate any number of pairs before triggering an iteration, in which computations are performed. Computations create or extend homology groups. The new homology information is propagated through the ontologies. The underlying idea is that if two concepts A and B are homologous, then one of the sub-concepts of A is probably homologous to one of the sub-concepts of B even if they have different names. Of note, validated homology contributes a significantly higher score than name similarity. Propagation is down weighted by the number of sub-concepts, to avoid generating many false positives (e.g. all the children of "whole body").

Evaluation of pairs, ordered by total score (base score + propagated score), and iteration, are repeated until the expert decides to terminate, or no more pairs are proposed. Compared to manual alignment of the ontologies, Homolonto reduces time considerably, with high sensitivity. Thus aligning the zebrafish (ZFIN; 2087 terms) and Xenopus (Xenbase; 480 terms) ontologies took one month by hand, but 2 days using Homolonto. The first 213 pairs proposed to the expert were valid at 80%, and contained 91% of all true positives.

To design homology relationships between several species, we merge the homology groups obtained by pair-wise alignment.

Finally, Homolonto generates an OBO [10] file containing the homology relationships. Bgee then parses this file to integrate the homologies into the database.

3   Mapping of the Developmental Ontologies

In relationship with the anatomical ontologies, Bgee uses for each species an ontology which describes its developmental stages, and links them using an *is_a* relationship by key states (e.g. embryo, hatching, larval).

To compare expression patterns, the comparisons have to be done both between homologous organs (see section 2), and at an equivalent developmental stage. But it is not possible to "simply" identify stages between species for which the state of the development is identical: organs do not develop at the same speed and with the same sequence, development is heterochronous (e.g. [11]).

A solution could be to identify, for each organ involved in a homology relationship, the different key states of their formation, and to design, organ by organ, equivalence relationship between these states in different species. This solution is difficult to implement, as it would imply manual definition for each organ separately, without any guiding principle in the data (i.e. we cannot use shared names and ontology structures as for anatomical homology).

Although there is no direct equivalence between the stages of two species because of heterochrony, it is instead possible to identify key events of development, common to all bilaterian animals. We have developed a small ontology of these common "metastages": embryo – including zygote, cleavage, blastula, gastrula, organogenesis –, post-embryonic development, adult. Then we have mapped the developmental stages of each species to these "metastages". This approach results in a loss of accuracy regarding the developmental ontologies, but allows to compare gene expression patterns taking into account the time dimension.

4   Integrating Heterogeneous Data on Anatomical and Developmental Ontologies

Integrating heterogeneous expression data is challenging, as it is difficult to compare the results of different types of techniques (e.g. ESTs, microarrays, *in situ* hybridizations) [12, 13], and even for a same type, to compare results between experiments (e.g. compare two microarray experiments made on different platforms). But as we want to be able to precisely describe expression patterns of genes, we need data as complete as possible. We also want to obtain data for all the species studied, and some techniques cannot be applied to all species, for instance *in situ* hybridizations on human. The information we want to collect is in which organs, and at which developmental stages, a gene is expressed. It means that for each experiment, we have to map the data to anatomical and developmental ontologies, and to apply statistical analyses, depending on the data type, to identify genes significantly expressed.

4.1   Mapping Expression Data to Ontologies

The main problem to map the data to ontologies is that annotations are often inconsistent between data sources: for instance, the description of the organs on which an experiment has been performed can be provided as free text, controlled vocabularies, or ontologies. Therefore, we have manually annotated each experiment stored in Bgee to determine the unique identifiers (ID) in the anatomical ontologies of the organs studied, and the ID of the developmental stages.

The granularity of the data is also highly variable. For instance, experiments can be reported on the organ "brain" or on the organ "forebrain", at the stage "embryo" or at the stage "free blastocyst". This is why ontologies are essential both for anatomy and for development: just listing the developmental stages would not have been sufficient.

4.2   Statistical Analyses

Bgee currently uses EST data from Unigene [14] and Affymetrix data retrieved from ArrayExpress [15]. For each data type, Bgee applies statistical tests to identify genes that are significantly expressed, with two levels of confidence: low and high.

For experiments based on tag counting, such as EST, SAGE, or MPSS, a statistical test [16] shows that a gene is expressed with a 95% confidence if 7 tags are mapped to this gene (the number of tags is statistically different from 0). So for EST data, we have considered a gene as expressed with a high confidence if an experiment has found at least 7 EST related to this gene, and with a low confidence from 1 to 6 EST.

Affymetrix data are measurements of fluorescence intensity. Labelled cDNAs prepared from samples are hybridized with oligonucleotide probes. All probes mapping to the same transcript constitute a probeset. Identifying genes significantly expressed consists in finding genes for which the signal of the probeset is significantly different from the background signal. This method is implemented by the MAS5 software [17]; based on these statistical analyses, probesets are flagged as "present", "marginal", or "absent". This allows us to classify genes expressed with a high confidence when their probeset is flagged as "present", and with a low confidence when "marginal". Although MAS5 classification is efficient [18], the estimation of the background signal can be biased depending on probe sequence affinity [19]. We are currently implementing another method of detection [19], which uses the gcRMA algorithm [20] to normalize the signal taking into account probe sequences, and uses a subset of weakly expressed probesets for estimating the background. A Wilcoxon test is then applied to compare the normalized signal of the probesets with the background signal. Genes will be considered expressed with a high confidence if the p-value is lower than 1%, and with a low confidence if the p-value is between 1 and 5 %.

Bgee will soon include *in situ* hybridization data. For data based on image analyses, statistical tests cannot be applied easily. Determining if a gene is expressed is usually done manually by an expert. A quality annotation can also be provided, summarizing the quality of the image, the hybridization, and the probes design. Such information is already present in several databases (e.g. ZFIN [8]), and Bgee will rely on them.

5   Database and Web-Interface of Bgee

The database of Bgee is developed with MySQL, and currently includes anatomical ontologies, developmental ontologies, and expression data for four species: human, mouse, zebrafish, and Xenopus:

- The anatomical ontologies come from eVoc [21] for human, Xspan [22] for human and mouse, MGD [23] for adult mouse, ZFIN [8] for zebrafish, and Xenbase [24] for Xenopus.
- EST data come from Unigene [14] and Affymetrix data from ArrayExpress [15]. *In situ* hybridization will be collected from specialized databases, as ZFIN or BGEM [25].
- Gene ontology [26] annotations and homology relationships between genes are recovered from Ensembl [27].
- Bgee currently includes a total of 104,881 genes. 51,277 have expression data, in 587 anatomical structures and 93 developmental stages.

The web interface of Bgee is developed in Java using the servlet container Tomcat, with a Model-View-Controller architecture. The user experience is improved by the use of AJAX technologies (Asynchronous Javascript And XML). The website of Bgee, available at http://bgee.unil.ch/, proposes several ways to easily retrieve or compare expression data:

- Querying the database: data can be queried for genes, gene families, anatomical structures, or developmental stages, based on their names, synonyms, abbreviations, identifiers, or descriptions.
- Browsing the ontologies: anatomical and developmental ontologies can be browsed as a tree structured view. Information about the genes expressed is displayed for each anatomical structure or developmental stages. The display of these expression data can be adjusted by selecting data type and data quality, or by entering a list of gene identifiers or of GO terms.

- Retrieving the expression pattern of a gene: the expression pattern of a gene is also displayed as a tree structured view of the organs where it is expressed, at the selected developmental stage. The data used to define the pattern can be modified by selecting the data type or data quality.
- Comparing the expression patterns of homologous genes: the expression patterns of a gene family can be compared choosing the species studied, and as for the ontology browsing, by selecting data type and quality, list of genes or of GO terms.

The homology relationships and developmental ontologies, both in OBO format, the Homolonto software and source code, and the Bgee database and source code, will soon be available on our website.

6 Conclusions

We have developed pipelines to integrate ontologies and expression data to Bgee, and automatically perform statistical analyses. We also have developed the Homolonto software to facilitate the design of homology relationships. We have paid great attention to make the Java code of Bgee easy to evolve, with a clean architecture and reusable components. We have thus implemented all the requirements to add more species and more data types into Bgee in the future. We plan to add in the short-term *in situ* hybridization data.

The multi-species computer coding and storage of expression patterns was an essential key to perform high throughput analyses. We will now be able to design analysis tools dedicated to the comparison of expression patterns, and to address open biological questions, such as the relationships between evolution of development and of gene expression, or the identification of candidate genes for diseases.

Acknowledgements. We thank Frederic Ricci for data annotation. Funding was provided by Etat de Vaud, the program Crescendo, the SIB, the Decrypthon program.

References

1. Carroll, S.: Endless Forms Most Beautiful: The New Science of Evo Devo and The Making of the Animal Kingdom. W. W. Norton & Company, New York (2005)
2. Yanai, I., Graur, D., et al.: Incongruent expression profiles between human and mouse orthologous genes suggest widespread neutral evolution of transcription control. Omics 8, 15–24 (2004)
3. Jordan, I.K., Marino-Ramirez, L., et al.: Evolutionary significance of gene expression divergence. Gene 345, 119–126 (2005)
4. Schlicht, M., Matysiak, B., et al.: Cross-species global and subset gene expression profiling identifies genes involved in prostate cancer response to selenium. BMC Genomics 5, 58 (2004)
5. Gu, Z., Nicolae, D., et al.: Rapid divergence in expression between duplicate genes inferred from microarray data. Trends Genet 18, 609–613 (2002)
6. Gu, X., Zhang, Z., et al.: Rapid evolution of expression and regulatory divergences after yeast gene duplication. Proc. Natl. Acad. Sci. USA 102, 707–712 (2005)

7. He, X., Zhang, J.: Rapid subfunctionalization accompanied by prolonged and substantial neofunctionalization in duplicate gene evolution. Genetics 169, 1157–1164 (2005)
8. Sprague, J., Clements, D., et al.: The Zebrafish Information Network (ZFIN): the zebrafish model organism database. Nucleic Acids Res. 31, 241–243 (2003)
9. Shvaiko, P., Euzenat, J.: Ontology Matching. Springer, Heidelberg (2007)
10. Smith, B., Ashburner, M., et al.: The OBO Foundry: coordinated evolution of ontologies to support biomedical data integration. Nat. Biotechnol. 25, 1251–1255 (2007)
11. Jeffery, J.E., Bininda-Emonds, O.R., et al.: A new technique for identifying sequence heterochrony. Syst. Biol. 54, 230–240 (2005)
12. Lee, C.K., Sunkin, S.M., et al.: Quantitative methods for genome-scale analysis of in situ hybridization and correlation with microarray data. Genome biology 9, R23 (2008)
13. Kuo, W.P., Liu, F., et al.: A sequence-oriented comparison of gene expression measurements across different hybridization-based technologies. Nat. Biotechnol. 24, 832–840 (2006)
14. Wheeler, D.L., Barrett, T., et al.: Database resources of the National Center for Biotechnology Information. Nucleic Acids Res. 36, 13–21 (2008)
15. Parkinson, H., Kapushesky, M., et al.: ArrayExpress–a public database of microarray experiments and gene expression profiles. Nucleic Acids Res. 35, 747–750 (2007)
16. Audic, S., Claverie, J.M.: The significance of digital gene expression profiles. Genome Res. 7, 986–995 (1997)
17. Liu, W.M., Mei, R., et al.: Analysis of high density expression microarrays with signed-rank call algorithms. Bioinformatics 18, 1593–1599 (2002)
18. Choe, S.E., Boutros, M., et al.: Preferred analysis methods for Affymetrix GeneChips revealed by a wholly defined control dataset. Genome biology 6, R16 (2005)
19. Schuster, E.F., Blanc, E., et al.: Correcting for sequence biases in present/absent calls. Genome biology 8, R125 (2007)
20. Wu, Z., Irizarry, R.A., et al.: A Model-Based Background Adjustment for Oligonucleotide Expression Arrays. Journal of the American Statistical Association 99, 909–917 (2004)
21. Kruger, A., Hofmann, O., et al.: Simplified ontologies allowing comparison of developmental mammalian gene expression. Genome biology 8, R229 (2007)
22. Aitken, S.: Formalizing concepts of species, sex and developmental stage in anatomical ontologies. Bioinformatics 21, 2773–2779 (2005)
23. Eppig, J.T., Blake, J.A., et al.: The mouse genome database (MGD): new features facilitating a model system. Nucleic Acids Res. 35, 630–637 (2007)
24. Bowes, J.B., Snyder, K.A., et al.: Xenbase: a Xenopus biology and genomics resource. Nucleic Acids Res. 36, 761–767 (2008)
25. Magdaleno, S., Jensen, P., et al.: BGEM: an in situ hybridization database of gene expression in the embryonic and adult mouse nervous system. PLoS Biol. 4, e86 (2006)
26. Ashburner, M., Ball, C.A., et al.: Gene ontology: tool for the unification of biology. The Gene Ontology Consortium. Nature genetics 25, 25–29 (2000)
27. Hubbard, T.J., Aken, B.L., et al.: Ensembl 2007. Nucleic Acids Res. 35, 610–617 (2007)

ENFIN - An Integrative Structure for Systems Biology

Florian Reisinger[1], Manuel Corpas[1], John Hancock[2], Henning Hermjakob[1], Ewan Birney[1], and Pascal Kahlem[1,*]

[1] EMBL - European Bioinformatics Institute, Wellcome Trust Genome Campus, Hinxton, Cambridge CB10 1SD, United Kingdom
[2] MRC Mammalian Genetics Unit, Harwell, Oxfordshire OX11 0RD, United Kingdom
`florian@ebi.ac.uk`, `corpas@ebi.ac.uk`, `j.hancock@har.mrc.ac.uk`,
`hhe@ebi.ac.uk`, `birney@ebi.ac.uk`, `pkahlem@ebi.ac.uk`
`http://www.enfin.org`

Abstract. Integration of biological data of various types and development of adapted bioinformatics tools represent critical objectives to enable research at the systems level. The European Network of Excellence ENFIN is engaged in developing an adapted infrastructure to connect databases, and platforms to enable both generation of new bioinformatics tools and experimental validation of computational predictions. With the aim of bridging the gap existing between standard wet laboratories and bioinformatics, the Network ENFIN runs integrative research projects, which require the development of adapted infrastructures that will be described in this paper: (i) a database infrastructure, EnCORE, appropriate for small laboratories, which can integrate public and local data, such as microarray data, protein-protein interaction data and pathway information; (ii) a registry of databases to serve as a reference of trusted databases; (iii) The database EnDICTION, to capture functional predictions generated by the computational analysis platform.

Keywords: ENFIN, Systems Biology, Integration.

1 Introduction

In the mid-twentieth century, measuring functions and behaviors of whole biological systems was the way to study biology because of the lack of knowledge and tools to address molecular questions [1, 2]. Classical systems biology led to the study of global phenomena such as growth, development or the influence of a given compound on the behavior of a cell, an organism, a population or even an ecosystem. With the development of molecular biology and the associated biotechnological methods, studies led first to the understanding of specific molecular events, for example the function of a single gene or a protein, the interactions between a few molecules or between domains of two molecules or the identification of the catalytic site of a given enzyme. The last decades have seen the adaptation of classical molecular biology methods to high-throughput scale, spanning from genome sequencing, gene

---

* Corresponding author.

A. Bairoch, S. Cohen-Boulakia, and C. Froidevaux (Eds.): DILS 2008, LNBI 5109, pp. 132–143, 2008.

expression analysis with microarrays, protein contents analysis with 2D gels electrophoresis and mass spectrometry, protein interactions analysis with yeast-2-hybrid technology [3] as well as large-scale screens using perturbation methods such as RNAi, chemicals or GFP-tagged protein expression. These large-scale approaches produce quantitative profiles of complete systems, requiring the use of adapted informatics tools for data sorting, comparison and modeling, amongst others. As a result of the technological development of high-throughput methods in biology, the scientific community is now presented with a growing collection of very heterogeneous types of data, whose integration should enable correlations with systems behaviors and functions.

2 Data Types and Standards

Qualitative biological information has been gathered in major widely used databases, containing catalogues of nucleotide sequences for multiple organisms (EMBL-BANK [4], GenBank [5]), genomes and the respective catalogues of genes (EnsEMBL [6]), proteins (UniProt [7]), protein identifications (PRIDE [8]), protein interactions (IntAct [9]), enzymes and their substrates (IntEnZ [10], Brenda [11]), carbohydrates, lipids or small chemical entities (ChEBI [12]), and many more. Supported by the use of the Internet, hundreds of laboratory-based databases share their contents online, most of them containing often very specific biological information. In addition, the design of these resources often corresponds to a particular need of the laboratory, making them difficult to exploit outside their original context. Recently, infrastructure developments have been driven by the need to open high-throughput biotechnologies to the wider community. Databases have been set up to collect and store large quantitative information resulting from high-throughput experiments, such as gene expression (ArrayExpress [13]), reaction kinetics (SABIO-RK [14]) or cellular phenotypes (MitoCheck [15]) amongst others. Alternative databases are used to store more elaborated information such as molecular reactions including metabolic and signal transduction pathways (Reactome [16], BioCyc [17], KEGG [18]). A variety of computational models of pathways is also available in the Biomodels database [19]. Scientific literature is also referenced in databases such as PubMed or CiteXplore, enabling single queries but also automated mining of the literature information such as iHOP [20]. The availability of qualitative and quantitative biological data in a digital format enables computed mining of the databases for various types of information. However, because of the use of various data formats and identifiers, the comparison and analysis of similar data types, such as genes or protein sets, originating from different sources is often laborious and impairs any further integration across different levels of analysis, such as protein function or gene/protein networks. Recent domain-specific recommendations for experimental data reporting comprise the minimum information about a microarray experiment (MIAME) [21], the minimum information about a proteomics experiment (MIAPE) [22], the minimum information required for reporting a molecular interaction experiment (MIMIx) [23], and the minimum information requested in the annotation of biochemical models (MIRIAM) [24]. Community standard data formats include the PSI MI format for molecular interactions [25], as well as BioPAX [26] and SBML

[27] for pathways representation. However, integration of minimum reporting requirements and community standard data formats across domains is still at an early stage, piloted by the MIBBI (mibbi.sf.net) and FuGe [28] initiatives, respectively (Table1). These recent initiatives towards defining standards to annotate biological information with dedicated ontologies should, if adopted by the scientific communities, enable the integration of newly produced datasets [29]. However, beyond the use of common standards to format individual datasets, there is a need for sophisticated informatics platforms to enable mining data across various sources, formats and types.

Table 1. Examples of databases referred-to in the text

Nucleotide Sequences	EMBL-BANK	www.ebi.ac.uk/embl
Genomes	Ensembl	www.ensembl.org
Proteins	UniProt	www.ebi.uniprot.org
Protein identification	PRIDE	www.ebi.ac.uk/pride
Enzymes	IntEnz	www.ebi.ac.uk/intenz
Enzymes	Brenda	www.brenda.uni-koeln.de
Small chemical entities	ChEBI	www.ebi.ac.uk/chebi
Curated human pathways	Reactome	www.reactome.org
Pathway/Genome Databases	BioCyc	biocyc.org/metacyc
Kyoto Encyclopedia of Genes and Genomes	KEGG	www.genome.jp/kegg
Gene expression datasets	Array Express	www.ebi.ac.uk/arrayexpress
Protein interaction	IntAct	www.ebi.ac.uk/intact
Reaction Kinetics	SABIO-RK	sabiork.villa-bosch.de
Cellular phenotypes	MitoCheck	www.mitocheck.org
Biological Models	BioModels	www.ebi.ac.uk/biomodels
Literature	CiteXplore	www.ebi.ac.uk/citexplore
Literature	PubMed	http://www.pubmed.gov/

3   The ENFIN Platform: Integrating Tools and Data

With a multidisciplinary consortium of 20 laboratories specialized in mathematics, computer sciences and biology, the European Network of Excellence ENFIN (Experimental Network for Functional Integration, www.enfin.org) develops methods and bioinformatics tools integrated in a common platform, divided mainly into four domains of analysis that are: prediction of protein function and protein interaction, network reconstruction and modeling. The bioinformatics tools developed in each domain of analysis are challenged on specific research projects. The Network maintains internally close collaboration between experimental and computational research, enabling a permanent cycling of experimental validation and improvement of computational prediction methods (figure 1). The toolbox EnSUITE, a provision of analysis tools developed and tested within ENFIN can be used as stand-alone products, but we aim at generating a series of web services that can serve as modules for workflow management software such as Taverna [30] or the one developed within ENFIN: EnVISION. Because of the diversity of research domains within ENFIN, bridging data types across different sources is a key issue which is addressed through

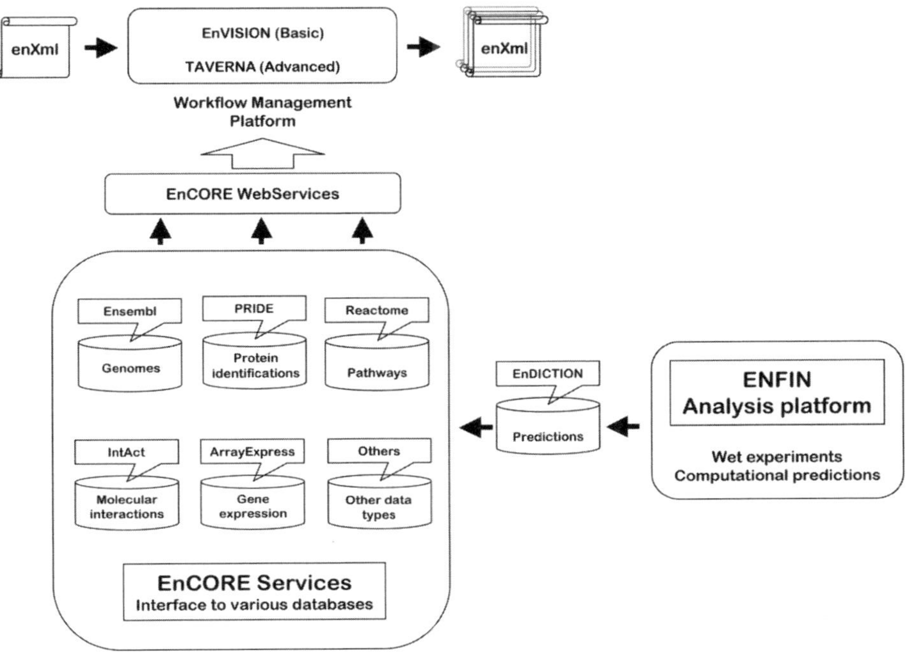

Fig. 1. The integrating structure of ENFIN. The EnCORE platform enables querying through series of different databases available as Webservices. The enXml file format allows the storage of the intermediate information collected during the query (documentation available at: www.enfin.org/encore-info). The analysis platform generates computational predictions, which will be integrated via the EnDICTION database.

EnCORE, a platform aimed at querying through various databases, thereby enabling data analysis and integration across different disciplines. Equipped with EnCORE, the ENFIN integrated platform should ultimately provide the user with a continuous suite of predictions tools and models intended for data interpretation beyond the classical one-tool/one-result analysis.

3.1   The EnCORE Platform

The need for open data access and data integration and combination across multiple data types has stimulated a variety of different approaches to solve this problem. Data warehouses, such as BioWarehouse [31] or SRS [32], integrate information from various resources into one single data storage system. While these approaches offer the advantage of a single point of entry and can perform very quick and complex queries on the data, they face the problem of frequently changing and growing data structures, which require a continuous adaptation and expansion of the system. Furthermore as a result of the data-gathering nature of the approach, the integrated data can never be as recent as the data in the original resource. Federated approaches avoid these problems by not duplicating the data, but rather allowing access to the distributed original data via a common interface or query language. EnCORE follows this federated approach, similarly to other projects such as Kleisli [33], Tambis [34] and others. The major advantage of the EnCORE system is, that it follows general community recommendations (Embrace; http://www.embracegrid.info) using open standards (SOAP, WSDL, XML, WS-I Basic Profile 1.0) and well supported technologies (JAX-WS, AJAX). Meanwhile EnCORE introduces as little new and specialized technologies as possible, such as sophisticated custom query languages and structures. Furthermore the EnCORE system is not limited to data retrieval only, but can also equally integrate analysis tools and algorithms. The bioinformatics community has recently started to provide further web services to address the need for language independent programmatic access across distributed data resources and analysis tools. However, most services are still focused on single entity operations and there is still a lack of interoperability, support of workflows, reproducibility or standard interfaces between web services. EnCORE also addresses these issues with its standard interface and data exchange format, together with the support of set-based operations and an in-document audit trail. The aim of the EnCORE project is to integrate an extensive list of database resources and analysis tools in a computationally accessible and extensible manner, facilitating automated data retrieval and processing with a particular focus on systems biology. Following the above requirements, the system is implemented as a growing set of modular web services following the Simple Object Access Protocol (SOAP). The use of the Web Service Definition Language (WSDL; http://www.ibm.com/developerworks/webservices/library/ws-whichwsdl) in the document/literal-wrapped style (in accordance with the WS-I Basic Profile 1.0 specifications (http://www.ws-i.org/Profiles/BasicProfile-1.0.html)) to specify the service interface guarantees interoperability. XML as data format ensures open access as well as platform and language independence. At the heart of EnCORE is enXml (http://www.enfin.org/encore/schema/documentation/enxml-v1.2.5-documentation.html), the XML schema defining the data exchange format between the web services, which provides a standardized, light-weight data structure for all EnCORE web services. It describes the core components of the ENFIN data model: *experiments*, *sets* and *molecules*. *Experiments* can include both wet lab experimental data as well as bioinformatics data transformations performed by EnCORE data retrieval or analysis services, and can comprise one or more input and result sets. *Sets* provide a convenient model to allow set-oriented bioinformatics operations on molecules or other sets. *Molecules* represent biological molecules such as proteins, which are traceable through multiple conversion steps, even when converting between data types in potentially ambiguous ways. EnCORE web services take enXml-schema-conform XML documents as input and produce modified documents as output by only adding a new *experiment* to its content describing the service procedure. This method preserves an audit trail, which is inseparable from the result and lies within the document itself. This standardization of the input and output values of the EnCORE web services also makes it very easy to chain the services into workflows by just passing-on the XML document from one service to another. It is up to the service to deal with the presented document content. Creating EnCORE workflows using workflow management tools such as Taverna becomes a simple task. The development of services has so far focused on databases hosted at the European Bioinformatics Institute (EBI), but the flexibility afforded by the generic enXml schema to integrate new data types will facilitate the addition of non-EBI private and public data sources in the future. EnCORE web services are currently available for PRIDE, IntAct, Reactome, UniProt, PICR [35], ArrayExpress, and an EnsEMBL based mapping between AffyMetrix probe set IDs and UniProt accessions. These services provide information on protein identifications, protein interactions, gene expression levels, pathway information, protein annotations, protein identifier mappings and probe set ID to protein identifier mapping, respectively.

The primary recipient will be EnSUITE, the ENFIN analysis layer. EnCORE web service interfaces for EnSUITE analysis tools are currently in develomment. All current services are implemented in Java using the Java JAX-WS framework, but the language independent nature of the system allows for implementations in other programming languages and sample applications using EnCORE web services are available in Java, Perl, Python and Taverna. The web application EnVISION (http://www.ebi.ac.uk/enfin-srv/envision) has been developed as a more end-user friendly interface to the EnCORE web services. It allows any number of services to be applied in any order to any enXml document, and for ease of usability it can create an initial enXml document from a given set of protein identifiers. It allows easily creating workflows involving various data retrieval and conversion steps without the need of additional applications or manual programming. EnVISION automatically converts the resulting enXml document into a human-readable HTML representation using an XSL Transformation script, which can also be used independently from the web application. The HTML form of the result also includes out-going links to the corresponding source databases (e.g. IntAct for protein interactions, Reactome for pathways) and is integrated into the start page of EnVISION, making it a light-weight one-page analysis tool. In the future, a new version of this interface, EnVISION II, will provide a more sophisticated Java Server Faces (JSF)-based web application with the aim to present the data contained in a enXml file in a clear, graphical way easily understandable for the user with outgoing links to the source databases for more detailed information. EnVISION II will also enable running queries in parallel, providing the possibility to start new queries even before earlier ones have returned any results.

As an example of the effectiveness of the EnCORE workflow system, we queried EnVISION with a small group of proteins (P22575, Q9UBU9, P06239) co-immunoprecipitated in an experiment described by Yoon et al. [36]. The proteins were shown by the IntAct database to interact with members of the T-cell signaling pathway, such as LAT and ZAP70. Data from Reactome further suggested a role in viral infectivity and life cycle and also an involvement of the proteins in mRNA processing. Data in PRIDE showed that many of proteins had been identified in samples from Jurkat T-cells. All of this information corroborates the observations of Yoon et al. who were investigating a novel Tip-associated protein (P22575) from *Herpesvirus saimiri,* which was shown to infect T lymphocytes and act as a mediator of T-cell transformation (Fig. 2).

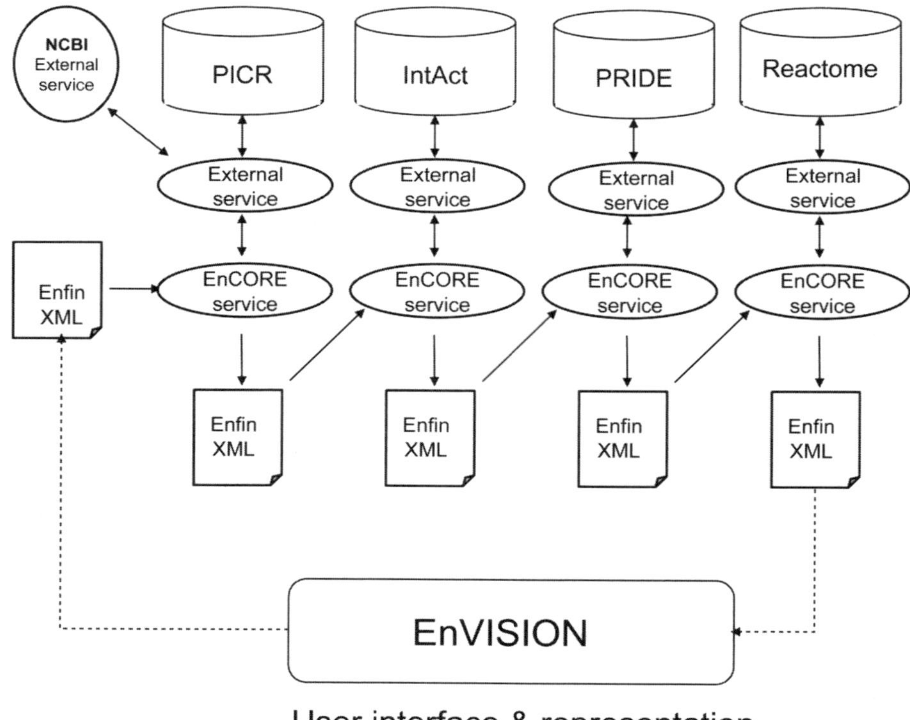

User interface & representation

Fig. 2. Example of workflow in EnVISION

3.2   Extending the Reach of EnCORE

The ENFIN registry of databases. A major issue for both bench scientists and the EnCORE team is the wide diversity of databases resources available in biology. To extend the scope of EnCORE and EnVISION beyond its core database set it will be necessary to collect a set of trusted databases relevant to systems biology that can be integrated into the system. The ENFIN consortium is therefore constructing a registry of systems biology databases that will act as a resource both to the EnCORE team and the systems biology community as a whole. Various attempts are being made to collect online lists of useful databases. A well-known example is the Nucleic Acids Research listing (http://www.oxfordjournals.org/nar/database/c) although this is relatively limited and broader in scope than the ENFIN Registry. An example of an online resource offering links to and information specifically about systems biology databases is Pathguide (http://www.pathguide.org/) [37]. This provides information on 240 databases related to systems biology, with a focus on proteins. The ENFIN Registry will lie between these resources in scope (for example only 19 of the resources currently on the Registry list are on Pathguide). Further, it aims at playing a distinct role by facilitating the integration of information in these databases rather than simply providing a useful user resource. In collaboration with another EU project, CASIMIR (Coordination and Sustainability of International Mouse Informatics Resources, www.casimir.org.uk) a specification is being developed for an information set that needs to be collected about databases to achieve this aim. This includes which (if any) ontologies are used for data annotation, and the types and specifications of any web services resources make available. The ENFIN registry, which will allow direct programmatic access, has the potential to serve as a repository of information that can be queried directly by EnVISION or other front ends. At the date of writing, the ENFIN registry list contains 60 databases and continues to be added to by members of the consortium. The database itself is under development in collaboration with a parallel database being developed by CASIMIR.

The EnDICTION database for predictions. A growing number of biological predictions are being generated within the "Analysis Platform" by the ENFIN consortium. We have chosen to start the development of this database with data of 3 types: protein interaction, protein function and sets of target genes. EnDICTION will enable the integration of such computational predictions with public databases available via the EnCORE platform. The data model in EnDICTION is based on the one used by the Reactome database (www.reactome.org), and consists of additional SQL tables designed to accommodate the 3 types of data proceeding from the ENFIN Analysis Platform. In EnDICTION, instances from a particular protein-protein interaction, list of target genes or functional attributes are referenced to their counterpart reaction in Reactome. An initial test case was performed where input of raw protein-protein interaction data for EnDICTION were obtained from an ENFIN partner and converted into the XML-derived standard PSI-MI. XML-formatted data were transformed into in-house Reactome-XML and were incorporated into the ENDICTION database. New relationships between entities were obtained from inference between already existing data and the newly incorporated. A specific field containing statistical measures of confidence for computational predictions is created for each batch of input data included in EnDICTION. At the end of the process, both curated Reactome and predicted ENFIN data are available seamlessly for the query. In the future, a graphical interface will allow overlaying reactions from Reactome with predicted functional interactions. In the context of the ENFIN project, EnDICTION will be adapted to store several annotated sets of predictions from different sources, compare them with the results of different experimental settings, or integrate results obtained using different technologies.

4 Perspectives

Enabling the integration of data from various sources, especially from small and middle-size laboratories is one of the concepts driving the ENFIN Network of Excellence, to further our understanding of biology through tight horizontal integration of data domains and vertical integration of databases, prediction systems, and experimental validation. Based on an open data integration platform, ENFIN development cycles will iteratively improve the quality of computational predictions through experimental validation, and will improve public resources through the provision of high quality systems biology models as well as supporting experimental findings.

Acknowledgements

We thank Antony Quinn and Sandra Orchard (European Bioinformatics Institute) for respectively initiating the EnVISION interface and preparing a test case of EnVISION. This work was supported by ENFIN, a Network of Excellence funded by the European Commission within its FP6 Programme, under the thematic area "Life sciences, genomics and biotechnology for health", contract number LSHG-CT-2005-518254.

References

1. Kacser, H., Burns, J.A.: The control of flux. Symp. Soc. Exp. Biol. 27, 65–104 (1973)
2. Von Bertalanffy, L.: The theory of open systems in physics and biology. Science 111, 23–29 (1950)
3. Legrain, P., Selig, L.: Genome-wide protein interaction maps using two-hybrid systems. FEBS Lett. 480, 32–36 (2000)
4. Kulikova, T., Akhtar, R., Aldebert, P., Althorpe, N., Andersson, M., Baldwin, A., Bates, K., Bhattacharyya, S., Bower, L., Browne, P., Castro, M., Cochrane, G., Duggan, K., Eberhardt, R., Faruque, N., Hoad, G., Kanz, C., Lee, C., Leinonen, R., Lin, Q., Lombard, V., Lopez, R., Lorenc, D., McWilliam, H., Mukherjee, G., Nardone, F., Pastor, M.P., Plaister, S., Sobhany, S., Stoehr, P., Vaughan, R., Wu, D., Zhu, W., Apweiler, R.: EMBL Nucleotide Sequence Database in 2006. Nucleic Acids Res. 35, 16–20 (2007)
5. Benson, D.A., Karsch-Mizrachi, I., Lipman, D.J., Ostell, J., Wheeler, D.L.: GenBank. Nucleic Acids Res. 35, 21–25 (2007)
6. Hubbard, T.J., Aken, B.L., Beal, K., Ballester, B., Caccamo, M., Chen, Y., Clarke, L., Coates, G., Cunningham, F., Cutts, T., Down, T., Dyer, S.C., Fitzgerald, S., Fernandez-Banet, J., Graf, S., Haider, S., Hammond, M., Herrero, J., Holland, R., Howe, K., Howe, K., Johnson, N., Kahari, A., Keefe, D., Kokocinski, F., Kulesha, E., Lawson, D., Longden, I., Melsopp, C., Megy, K., Meidl, P., Ouverdin, B., Parker, A., Prlic, A., Rice, S., Rios, D., Schuster, M., Sealy, I., Severin, J., Slater, G., Smedley, D., Spudich, G., Trevanion, S., Vilella, A., Vogel, J., White, S., Wood, M., Cox, T., Curwen, V., Durbin, R., Fernandez-Suarez, X.M., Flicek, P., Kasprzyk, A., Proctor, G., Searle, S., Smith, J., Ureta-Vidal, A., Birney, E.: Ensembl 2007. Nucleic Acids Res. 35, 610–617 (2007)
7. The UniProt Consortium: The Universal Protein Resource (UniProt). Nucleic Acids Res. 35, 193–197 (2007)
8. Jones, P., Cote, R.G., Cho, S.Y., Klie, S., Martens, L., Quinn, A.F., Thorneycroft, D., Hermjakob, H.: PRIDE: new developments and new datasets. Nucleic Acids Res. 36, 878–883 (2008)
9. Kerrien, S., Alam-Faruque, Y., Aranda, B., Bancarz, I., Bridge, A., Derow, C., Dimmer, E., Feuermann, M., Friedrichsen, A., Huntley, R., Kohler, C., Khadake, J., Leroy, C., Liban, A., Lieftink, C., Montecchi-Palazzi, L., Orchard, S., Risse, J., Robbe, K., Roechert, B., Thorneycroft, D., Zhang, Y., Apweiler, R., Hermjakob, H.: IntAct–open source resource for molecular interaction data. Nucleic Acids Res 35, 561–565 (2007)
10. Fleischmann, A., Darsow, M., Degtyarenko, K., Fleischmann, W., Boyce, S., Axelsen, K.B., Bairoch, A., Schomburg, D., Tipton, K.F., Apweiler, R.: IntEnz, the integrated relational enzyme database. Nucleic Acids Res. 32, 434–437 (2004)
11. Schomburg, I., Chang, A., Ebeling, C., Gremse, M., Heldt, C., Huhn, G., Schomburg, D.: BRENDA, the enzyme database: updates and major new developments. Nucleic Acids Res. 32, 431–433 (2004)

12. Degtyarenko, K., de Matos, P., Ennis, M., Hastings, J., Zbinden, M., McNaught, A., Alcantara, R., Darsow, M., Guedj, M., Ashburner, M.: ChEBI: a database and ontology for chemical entities of biological interest. Nucleic Acids Res. 36, 344–350 (2008)

13. Parkinson, H., Kapushesky, M., Shojatalab, M., Abeygunawardena, N., Coulson, R., Farne, A., Holloway, E., Kolesnykov, N., Lilja, P., Lukk, M., Mani, R., Rayner, T., Sharma, A., William, E., Sarkans, U., Brazma, A.: ArrayExpress–a public database of microarray experiments and gene expression profiles. Nucleic Acids Res. 35, 747–750 (2007)

14. Rojas, I., Golebiewski, M., Kania, R., Krebs, O., Mir, S., Weidemann, A., Wittig, U.: Storing and annotating of kinetic data. Silico Biol. 7, 37–44 (2007)

15. Erfle, H., Neumann, B., Liebel, U., Rogers, P., Held. M., Walter, T., Ellenberg, J., Pepperkok, R.: Reverse transfection on cell arrays for high content screening microscopy. Nat. Protoc. 2, 392–399 (2007)

16. Vastrik, I., D'Eustachio, P., Schmidt, E., Joshi-Tope, G., Gopinath, G., Croft, D., de Bono, B., Gillespie, M., Jassal, B., Lewis, S., Matthews, L., Wu, G., Birney, E., Stein, L.: Reactome: a knowledge base of biologic pathways and processes. Genome Biol. 8, R39 (2007)

17. Karp, P.D., Ouzounis, C.A., Moore-Kochlacs, C., Goldovsky, L., Kaipa, P., Ahren, D., Tsoka, S., Darzentas, N., Kunin, V., Lopez-Bigas, N.: Expansion of the BioCyc collection of pathway/genome databases to 160 genomes. Nucleic Acids Res. 33, 6083–6089 (2005)

18. Kanehisa, M., Goto, S., Hattori, M., Aoki-Kinoshita, K.F., Itoh, M., Kawashima, S., Katayama, T., Araki, M., Hirakawa, M.: From genomics to chemical genomics: new developments in KEGG. Nucleic Acids Res. 34, 354–357 (2006)

19. Le Novere, N., Bornstein, B., Broicher, A., Courtot, M., Donizelli, M., Dharuri, H., Li, L., Sauro, H., Schilstra, M., Shapiro, B., Snoep, J.L., Hucka, M.: BioModels Database: a free, centralized database of curated, published, quantitative kinetic models of biochemical and cellular systems. Nucleic Acids Res. 34, 689–691 (2006)

20. Fernandez, J.M., Hoffmann, R., Valencia, A.: iHOP web services. Nucleic Acids Res. (2007)

21. Brazma, A., Hingamp, P., Quackenbush, J., Sherlock, G., Spellman, P., Stoeckert, C., Aach, J., Ansorge, W., Ball, C.A., Causton, H.C., Gaasterland, T., Glenisson, P., Holstege, F.C., Kim, I.F., Markowitz, V., Matese, J.C., Parkinson, H., Robinson, A., Sarkans, U., Schulze-Kremer, S., Stewart, J., Taylor, R., Vilo, J., Vingron, M.: Minimum information about a microarray experiment (MIAME)-toward standards for microarray data. Nat. Genet. 29, 365–371 (2001)

22. Taylor, C.F., Paton, N.W., Lilley, K.S., Binz, P.A., Julian Jr., R.K., Jones, A.R., Zhu, W., Apweiler, R., Aebersold, R., Deutsch, E.W., Dunn, M.J., Heck, A.J., Leitner, A., Macht, M., Mann, M., Martens, L., Neubert, T.A., Patterson, S.D., Ping, P., Seymour, S.L., Souda, P., Tsugita, A., Vandekerckhove, J., Vondriska, T.M., Whitelegge, J.P., Wilkins, M.R., Xenarios, I., Yates III, J.R., Hermjakob, H.: The minimum information about a proteomics experiment (MIAPE). Nat. Biotechnol. 25, 887–893 (2007)

23. Orchard, S., Salwinski, L., Kerrien, S., Montecchi-Palazzi, L., Oesterheld, M., Stumpflen, V., Ceol, A., Chatr-aryamontri, A., Armstrong, J., Woollard, P., Salama, J.J., Moore, S., Wojcik, J., Bader, G.D., Vidal, M., Cusick, M.E., Gerstein, M., Gavin, A.C., Superti-Furga, G., Greenblatt, J., Bader, J., Uetz, P., Tyers, M., Legrain, P., Fields, S., Mulder, N., Gilson, M., Niepmann, M., Burgoon, L., De Las Rivas, J., Prieto, C., Perreau, V.M., Hogue, C., Mewes, H.W., Apweiler, R., Xenarios, I., Eisenberg, D., Cesareni, G., Hermjakob, H.: The minimum information required for reporting a molecular interaction experiment (MIMIx). Nat. Biotechnol. 25, 894–898 (2007)

24. Le Novere, N., Finney, A., Hucka, M., Bhalla, U.S., Campagne, F., Collado-Vides, J., Crampin, E.J., Halstead, M., Klipp, E., Mendes, P., Nielsen, P., Sauro, H., Shapiro, B., Snoep, J.L., Spence, H.D., Wanner, B.L.: Minimum information requested in the annotation of biochemical models (MIRIAM). Nat. Biotechnol. 23, 1509–1515 (2005)
25. Hermjakob, H., Montecchi-Palazzi, L., Bader, G., Wojcik, J., Salwinski, L., Ceol, A., Moore, S., Orchard, S., Sarkans, U., von Mering, C., Roechert, B., Poux, S., Jung, E., Mersch, H., Kersey, P., Lappe, M., Li, Y., Zeng, R., Rana, D., Nikolski, M., Husi, H., Brun, C., Shanker, K., Grant, S.G., Sander, C., Bork, P., Zhu, W., Pandey, A., Brazma, A., Jacq, B., Vidal, M., Sherman, D., Legrain, P., Cesareni, G., Xenarios, I., Eisenberg, D., Steipe, B., Hogue, C., Apweiler, R.: The HUPO PSI's molecular interaction format–a community standard for the representation of protein interaction data. Nat. Biotechnol. 22, 177–183 (2004)
26. Luciano, J.S.: PAX of mind for pathway researchers. Drug Discov. Today 10, 937–942 (2005)
27. Hucka, M., Finney, A., Sauro, H.M., Bolouri, H., Doyle, J.C., Kitano, H., Arkin, A.P., Bornstein, B.J., Bray, D., Cornish-Bowden, A., Cuellar, A.A., Dronov, S., Gilles, E.D., Ginkel, M., Gor, V., Goryanin II, H.W.J., Hodgman, T.C., Hofmeyr, J.H., Hunter, P.J., Juty, N.S., Kasberger, J.L., Kremling, A., Kummer, U., Le Novere, N., Loew, L.M., Lucio, D., Mendes, P., Minch, E., Mjolsness, E.D., Nakayama, Y., Nelson, M.R., Nielsen, P.F., Sakurada, T., Schaff, J.C., Shapiro, B.E., Shimizu, T.S., Spence, H.D., Stelling, J., Takahashi, K., Tomita, M., Wagner, J., Wang, J.: The systems biology markup language (SBML): a medium for representation and exchange of biochemical network models. Bioinformatics 19, 524–531 (2003)
28. Jones, A.R., Miller, M., Aebersold, R., Apweiler, R., Ball, C.A., Brazma, A., Degreef, J., Hardy, N., Hermjakob, H., Hubbard, S.J., Hussey, P., Igra, M., Jenkins, H., Julian Jr., R.K., Laursen, K., Oliver, S.G., Paton, N.W., Sansone, S.A., Sarkans, U., Stoeckert Jr., C.J., Taylor, C.F., Whetzel, P.L., White, J.A., Spellman, P., Pizarro, A.: The Functional Genomics Experiment model (FuGE): an extensible framework for standards in functional genomics. Nat. Biotechnol. 25, 1127–1133 (2007)
29. Ball, C.A.: Are we stuck in the standards? Nat. Biotechnol. 24, 1374–1376 (2006)
30. Hull, D., Wolstencroft, K., Stevens, R., Goble, C., Pocock, M.R., Li, P., Oinn, T.: Taverna: a tool for building and running workflows of services. Nucleic Acids Res. 34, 729–732 (2006)
31. Lee, T.J., Pouliot, Y., Wagner, V., Gupta, P., Stringer-Calvert, D.W., Tenenbaum, J.D., Karp, P.D.: BioWarehouse: a bioinformatics database warehouse toolkit. BMC Bioinformatics 7, 170 (2006)
32. Zdobnov, E.M., Lopez, R., Apweiler, R., Etzold, T.: The EBI SRS server-new features. Bioinformatics 18, 1149–1150 (2002)
33. Lin, K., Ting, A.E., Wang, J., Wong, L.: Hunting TPR Domains Using Kleisli. In: Genome Inform Ser Workshop Genome Inform, vol. 9, pp. 173–182 (1998)
34. Stevens, R., Baker, P., Bechhofer, S., Ng, G., Jacoby, A., Paton, N.W., Goble, C.A., Brass, A.: TAMBIS: transparent access to multiple bioinformatics information sources. Bioinformatics 16, 184–185 (2000)
35. Cote, R.G., Jones, P., Martens, L., Kerrien, S., Reisinger, F., Lin, Q., Leinonen, R., Apweiler, R., Hermjakob, H.: The Protein Identifier Cross-Referencing (PICR) service: reconciling protein identifiers across multiple source databases. BMC Bioinformatics 8, 401 (2007)

36. Yoon, D.W., Lee, H., Seol, W., DeMaria, M., Rosenzweig, M., Jung, J.U.: Tap: a novel cellular protein that interacts with tip of herpesvirus saimiri and induces lymphocyte aggregation. Immunity 6, 571–582 (1997)
37. Bader, G.D., Cary, M.P., Sander, C.: Pathguide: a pathway resource list. Nucleic Acids Res. 34, 504–506 (2006)

A System for Ontology-Based Annotation of Biomedical Data

Clement Jonquet, Mark A. Musen, and Nigam Shah

Stanford Center for Biomedical Informatics Research
Stanford University School of Medicine
Medical School Office Building, Room X-215
251 Campus Drive, Stanford, CA 94305-5479 USA
{jonquet,musen,nigam}@stanford.edu

Abstract. We present a system for ontology based annotation and indexing of biomedical data; the key functionality of this system is to provide a service that enables users to locate biomedical data resources related to particular ontology concepts. The system's indexing workflow processes the text metadata of diverse resource elements such as gene expression data sets, descriptions of radiology images, clinical-trial reports, and PubMed article abstracts to annotate and index them with concepts from appropriate ontologies. The system enables researchers to search biomedical data sources using ontology concepts. What distinguishes this work from other biomedical search tools is:(i) the use of ontology semantics to expand the initial set of annotations automatically generated by a concept recognition tool; (ii) the unique ability to use almost all publicly available biomedical ontologies in the indexing workflow; (iii) the ability to provide the user with integrated results from different biomedical resource in one place. We discuss the system architecture as well as our experiences during its prototype implementation (http://www.bioontology.org/tools.html).

Keywords: ontology-based annotation, biomedical data integration, biomedical ontologies, semantic expansion, concept recognition.

1 Introduction

The emergence of information and communication technologies has drastically changed biomedical scientific processes. Experimental data and results today are easy to share and repurpose thanks to the Web and public application programming interfaces (APIs) enabling connection to databases containing such information. As a consequence, the variety of biomedical data available in the public domain is now very diverse and ranges from genomic-level high-throughput data to molecular-imaging studies to published research articles. The paradox of such an expansion is that biomedical researchers now face the problem of extracting the specific data they need. Measures must be taken to prevent this problem from worsening as data

A. Bairoch, S. Cohen-Boulakia, and C. Froidevaux (Eds.): DILS 2008, LNBI 5109, pp. 144–152, 2008.

repositories grow fast[1]. Biomedical researchers have turned to ontologies and terminologies to describe their data and turn it into structured and formalized knowledge. For instance, the Gene Ontology[2] (GO) is widely used to describe the molecular functions, cellular location and biological processes of gene products as well as integrate these descriptions across several databases.

However, most publicly available biomedical data are unstructured and rarely described with ontology concepts available in the domains. This wealth of publicly accessible biomedical data is beginning to enable cross-cutting integrative translational bioinformatics studies [1][2]. In order to develop integrative translational bioinformatics approaches to interpret these datasets, there is a strong and pressing need to be able to identify all experiments that study a particular disease. A key query dimension for such integrative studies is the sample, along with a gene or protein name. As a result, besides queries that identify all genes that have a function X – which can be reliably answered using GO – we need to conduct queries that find all samples/experiments that study a particular disease and/or the effect of an experimental agent. However, translational discoveries that could be made by mining biomedical resources are hampered because they lack standard terminologies and ontologies to describe their elements (i.e., diagnoses, diseases, samples, and experimental conditions). For example, a researcher studying the allelic variations in a gene would want to know all the pathways that are affected by that gene, the drugs whose effects could be modulated by the allelic variations in the gene, and any disease that could be caused by the gene, and the clinical trials that have studied drugs or diseases related to that gene. The knowledge needed to study such questions is available in public biomedical resources; the problem is finding that information.

The challenge is to create consistent terminology labels for each element in the public resources that would allow the identification of all elements that relate to the same type at a given level of granularity. (e.g., *All carcinoma* samples versus *all Adenocarcinoma in situ of prostate* samples, where the former is at a coarser level of detail). These resource elements range from experimental data sets in repositories, to records of disease associations of gene products in mutation databases, to entries of clinical-trial descriptions, to published papers, and so on. One mechanism of achieving this objective is to map the text metadata describing the diagnoses, pathological state and experimental agents applied to a particular sample to ontology concepts allowing us to formulate refined or coarse search criteria. Creating ontology-based annotations from these resource elements metadata will enable end users to formulate flexible searches for biomedical data [3][4][5][6][7]. Therefore, the key challenge is to automatically and consistently annotate the biomedical data resource elements to identify the biomedical concepts to which they relate.

In this paper, we present a system for ontology-based annotation, which enables users to locate biomedical data related to particular ontology concepts in the BioPortal[3] ontology repository. The system's indexing workflow processes the text

---

[1] For example, in February 2007, the Gene Expression Omnibus (GEO) had 369 data sets; in the March 2007 release, the number of data sets increased to about 1500 and is now, in February 2008, around 2085 data sets.

[2] www.geneontology.org/

[3] www.bioontology.org/tools/portal/bioportal.html

metadata of several biomedical resource elements to annotate (or tag) them with concepts from appropriate ontologies and create an index to access these elements. As described in the following sections, the tagging is done with a concept recognition tool and the final index takes into accounts the ontology semantics that link concepts to one another (e.g., is_a relation). Our system creates an ontology-based index that can be used by existing search engines (such as Entrez, BioNavigator) to retrieve results that are complementary to the ones found with keyword based approaches. What distinguishes our system is: (i) the use of ontology semantics (ii) the ability to use almost all publicly available biomedical terminologies such as the Unified Medical Language System (UMLS) ontologies as well as Open Biomedical Ontologies, in the indexing workflow; (iii) the ability to provide the user with integrated results from different biomedical resource in one place. In the rest of the paper, Section 2 introduces the system architecture Section 3 gives an example on a GEO dataset. Section 4 presents our implemented prototype and the integration of its results in BioPortal. Section 5 concludes.

2   System Architecture

In this section we describe the system architecture consisting of different levels (Fig. 1). At the *resource level*, public biomedical resources (such as GEO and PubMed) are composed of elements that represent an abstraction for the unit of storage in theses databases. An element is identifiable and can be linked by a specific URL/URI (id), and it has a structure that defines the metadata contexts for the element (title, description, abstract, and so on). Our system retrieves[4] and downloads (through specific access tools) the element text metadata from resources, and keeps a track from both the original metadata context and element id. At the *annotation level*, the system uses a concept recognition tool called mgrep (developed by Univ. of Michigan) to annotate (or tag) resource elements with terms from a dictionary. The dictionary is constructed by including all the concept names and synonyms from a set of ontologies available at the ontology level. The annotation process is context aware, and keeps track of the context (such as title, description) from which the annotation was derived. The results are stored as annotation tables. An annotation table contains information such as *"element E was annotated with concept T in context C"*.

At the *index level*, a global index combines all the annotation tables and indexes annotations according to ontology concepts. The index contains information such as: *"Concept T annotates elements E1, E2, ..."*.

The system also uses relations provided at the *ontology level* to expand the annotations. This is the first step of the semantic expansion. For example, using the is_a ontology relation, for each annotation, we create additional transitive closure annotations according to the parent–child relationships subsumed by the original concept. For instance, if a resource element such as a GEO protein expression study is annotated with a concept from the ontology National Cancer Institute Thesaurus (NCIT), e.g., *pheochromocytoma*, then a researcher can query for *retroperitoneal neoplasms* and find data sets related to *pheochromocytoma*. The NCIT provides the

---

[4] We use public API such as Web Services or structured XML documents.

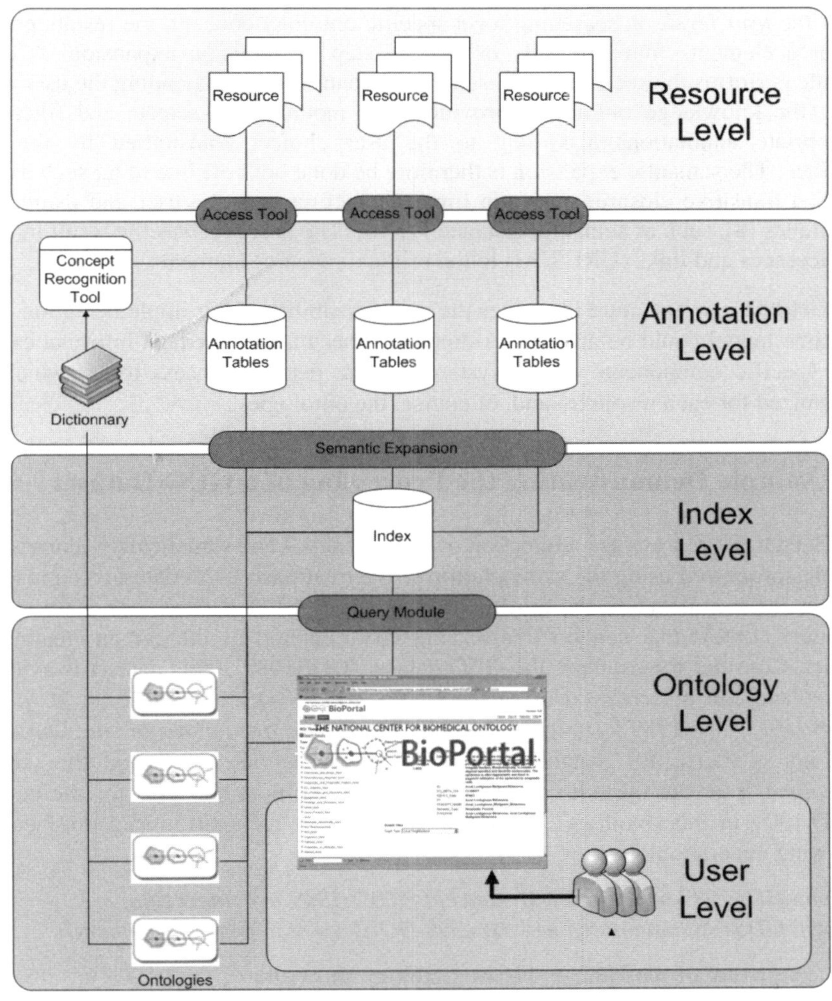

Fig. 1. The system architecture comprising of different levels. See main text for details.

knowledge that *pheochromocytoma* is_a *retroperitoneal neoplasms*. This first step is done offline because, processing the transitive closure is very time consuming – even if we use a pre-computed hierarchy – and will result in prolonged response times for the users. This use case is similar, in principle, to query expansion done by search engine like Entrez; however, Entrez does not use ontologies, therefore, there exists *pheochromocytoma* related GEO data sets, but none show up on searching for retroperitoneal neoplasms in Entrez. In our system, however, a researcher could search for *retroperitoneal neoplasms* and find the relevant samples [1].

At the *user level*, on searching for a specific ontology concept, the results provide resource elements found directly or via the step of semantic expansion. A query module performs the second step of semantic expansion i.e., expanding the user query using the knowledge ontologies provide. This module also selects and filters the appropriate annotations according to the user choices transmitted by the user interface. The semantic expansion is therefore be done both off line (e.g., such as with the is_a transitive closure) or at run time, interacting with the user and using other techniques [8], such as semantic distance [9][10]. The user receives the result in terms of references and links (URL/URI) to the original resource elements.

Remark: This architecture illustrates the generalizability of our implementation. Note the same model could be applied for domains other than biomedical informatics. The only specific components of the system are the resource access tools (which are customized for each resource) and, of course, the ontologies.

3   Example Demonstrating the Processing of a GEO Dataset

A GEO dataset represents a collection of biologically – and statistically – comparable samples processed using the same platform. We treat each GEO dataset as a resource element whose metadata we aim to process. Each GEO dataset, has a title and a summary context that contain free text metadata entered by the person creating the dataset. Consider for example the GEO dataset 'GDS1989'. This dataset is available online[5] and can be retrieved using the EUtils API.[6] GDS1989's title is: *Melanoma progression*. GDS1989's summary contains the phrase: *melanoma in situ*. Our set of ontology contains, for instance, the Human disease ontology,[7] and the concept *Melanoma* is in our system's dictionary as it is one possible term for the concept DOID:1909 in this ontology. Therefore, our concept recognition tool produces the following annotations:[8]

Element GDS1989 annotated with concept DOID:1909 in context title;
Element GDS1989 annotated with concept DOID:1909 in context summary;

The structure of the Human disease ontology shows that DOID:1909 has 36 direct or indirect parents such as for instance DOID:169, *Neuroendocrine Tumors* and DOID:4, *Disease*, therefore the transitive closure on the is_a relation generates, for instance, the following annotations:

Element GDS1989 annotated with concept DOID: 169 with closure;
Element GDS1989 annotated with concept DOID:4 with closure;

Searching for "melanoma" in BioPortal returns 109 matches[9] in the Human disease ontology including concept DOID1909. The user can access the 13 ArrayExpress

---

[5] www.ncbi.nlm.nih.gov/projects/geo/gds/gds_browse.cgi?gds=1989

[6] www.ncbi.nlm.nih.gov/entrez/query/static/eutils_help.html

[7] http://diseaseontology.sourceforge.net/

[8] Note these two annotations involve only one annotating concept.

[9] BioPortal uses an Apache Lucene index provided by LexGrid (http://informatics.mayo.edu) to find the query related ontology concepts.

experiments, or the 673 clinical trials, or the 960 articles in PubMed and the 10 GEO datasets related to that concept.

4   Integration with NCBO BioPortal

The National Center for Biomedical Ontology (NCBO) [11] develops and maintains a Web application called BioPortal to access biomedical ontologies. This library contains a large collection of ontologies, such as GO, NCIT, International Classification of Diseases (ICD), in different formats (OBO, OWL, etc.). Users can browse and search this repository of ontologies both online and via a Web services API.

We have implemented the first prototype of the system as presented in section 2. We have written a set of Java access tools to access five resource databases. Resources processed and the numbers of annotations currently available in our system index are presented in Table 1. A public representational state transfer (REST) services API [12] is available to query the annotation index and returns XML documents describing the annotations. We have used this API to integrate the system with BioPortal as illustrated by Fig. 2.

In our prototype, we have processed: (1) high-throughput gene-expression data sets from GEO and Array Express, (2) clinical-trial descriptions from Clinicaltrials.gov, (3) captions of images from ARRS Goldminer, and (4) abstracts of articles published in PubMed. Table 1 shows both the current number of elements annotated and the number of annotations created from each resource that we have processed. Our prototype uses 48 different biomedical ontologies that give us a dictionary of 793681 unique concepts and 2130700 terms. As a result of using such a large number of terms, our system provides annotations for 99% of our subset of PubMed, and 100% of the other processed resources. The average number of annotating concepts is between 359 and 769 per element, with an average of 27% of these annotations being direct. In the current prototype, concept recognition is done using a concept recognition tool developed by National Center for Integrative Biomedical Informatics

Table 1. Number of elements annotated from each resource in the current prototype

Resource	Number of elements	Resource local size (Mb)	Number of direct annotations (mgrep results)	Total number of 'useful'[1] annotations	Average number of annotating concepts
PubMed (subset) www.ncbi.nlm.nih.gov/PubMed/	1050000	146.1	30822190	174840027	160
ArrayExpress www.ebi.ac.uk/arrayexpress/	3371	3.6	502122	1849224	525
ClinicalTrials.gov http://clinicaltrials.gov/	50303	99	16108580	48796501	824
Gene Expression Omnibus www.ncbi.nlm.nih.gov/geo/	2085	0.7	165539	772608	359
ARRS GoldMiner (subset) http://goldminer.arrs.org	1155	0.5	134229	662687	564
TOTAL	1106914	249.9	47732660	226921047	(avg)486.4

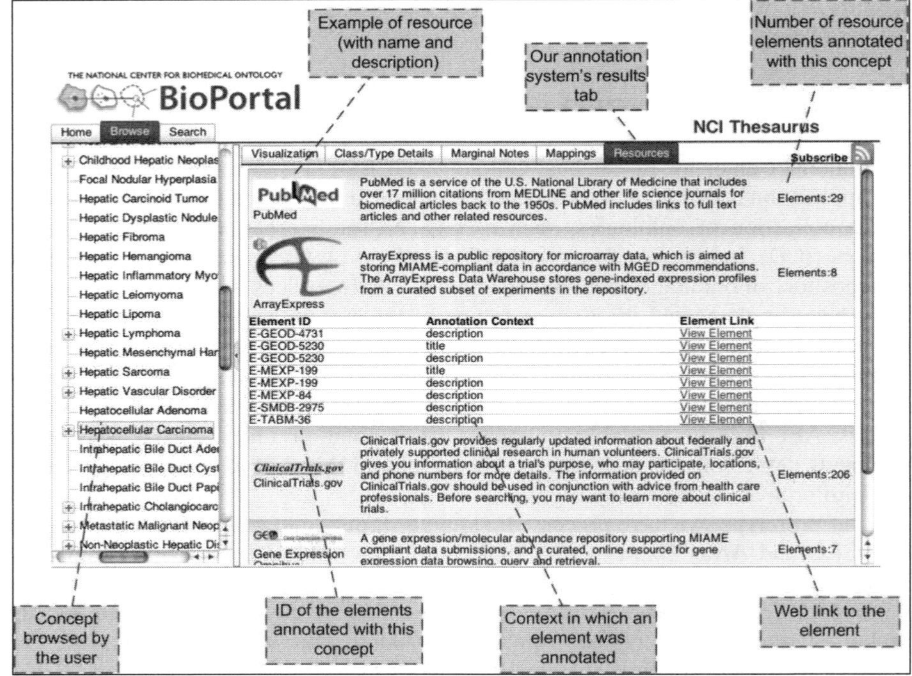

Fig. 2. User interface within BioPortal. In this view, a user browsing the NCIT in BioPortal, can select an ontology concept (in this case, *Hepatocellular carcinoma*) and see immediately the numbers of online resource elements that relate directly to that concept (and the concepts that it subsumes). The interface allows the user to directly access the original elements that are associated with *Hepatocellular carcinoma* for each of the indexed resources.

(NCIBI) called *mgrep*. [10]We rely on this tool which reported a very high degree of accuracy (over 95%) in recognizing disease names [13]. The prototype design of the annotation level is such that we can plug-in other concept recognizers. The prototype is available online http://alpha.bioontology.org/.

5   Conclusion

In this paper, we have described the prototype implementation of an ontology-based annotation system. The system's objective is to annotate (offline) a large number of biomedical resources and to provide an index up to date of annotated resources elements. We use ontologies (and not simply terminologies) both for annotation as

---

[10] We have conducted a comparative evaluation of this tool with the gold standard in the biomedical community, MetaMap [14]. It has a higher precision in recognizing concepts, and it is more scalable as well as open to outside dictionary (not tied to the UMLS structure as MetaMap is.).

well as semantic expansion of the annotations. The NCBO hosts one of the largest library of biomedical ontologies and our system allows a user to search for various biomedical data related to a specific ontology concepts in one place; greatly enhancing the value of the ontology repository. Our system can process text metadata of gene-expression data sets, descriptions of radiology images, clinical-trial reports, as well as abstracts of PubMed articles to annotate them automatically with concepts from appropriate ontologies. It promotes biomedical translational research by enabling users to locate relevant biological data sets and to integrate them with clinical data to bridge the bench-to-bedside gap.

We believe that as we expand the system with additional ontologies and process additional biomedical resources, we will serve an even wider user population, broadening the reach and impact of the NCBO in enabling translational research.

Acknowledgements

This work is supported by the National Center for Biomedical Computing (NCBC) National Institute of Health roadmap initiative; NIH grant U54 HG004028. We also acknowledge assistance of Manhong Dai and Fan Meng at University of Michigan as well as Chuck Kahn for the access to the Goldminer resource.

References

[1] Butte, A.J., Kohane, I.: Creation and implications of a phenome-genome network. Nature Biotechnology 24(1), 55–62 (2006)

[2] Butte, A., Chen, R.: Finding disease-related genomic experiments within an international repository: first steps in translational bioinformatics. In: American Medical Informatics Association Annual Symposium, AMIA 2006, Washington DC, USA, p. 106 (2006)

[3] Spasic, I., et al.: Text mining and ontologies in biomedicine: making sense of raw text. Brief Bioinformatics 6(3), 239 (2005)

[4] Moskovitch, R., Martins, S.B., Behiri, E., Weiss, A., Shahar, Y.: A Comparative Evaluation of Full-text, Concept-based, and Context-sensitive Search. American Medical Informatics Association 14(2), 164–174 (2007)

[5] Sneiderman, C.A., et al.: Knowledge-based Methods to Help Clinicians Find Answers in Medline. American Medical Informatics Association 14(6), 772–780 (2007)

[6] Shah, N.H., Rubin, D.L., Supekar, K.S., Musen, M.A.: Ontology-based Annotation and Query of Tissue Microarray Data. In: American Medical Informatics Association Annual Symposium, AMIA 2006, Washington DC, USA, pp. 709–713 (2006)

[7] Khelif, K., Dieng-Kuntz, R., Barbry, P.: An ontology-based approach to support text mining and information retrieval in the biological domain. Universal Computer Science 13(12), 1881–1907 (2007), Special Issue on Ontologies and their Applications

[8] Bhogal, J., Macfarlane, A., Smith, P.: A review of ontology based query expansion. Information Processing and Management 43, 866–886 (2007)

[9] Lee, W.J., Raschid, L., Srinivasan, P., Shah, N., Rubin, D., Noy, N.: Using Annotations from Controlled Vocabularies to Find Meaningful Associations. In: Cohen-Boulakia, S., Tannen, V. (eds.) DILS 2007. LNCS (LNBI), vol. 4544, pp. 264–279. Springer, Heidelberg (2007)

[10] Caviedesa, J.E., Cimino, J.J.: Towards the development of a conceptual distance metric for the UMLS. Biomedical Informatics 37(2), 77–85 (2004)

[11] Ashburner, M., Sim, I., Hute, C.G., Solbrig, H., Storey, M.A., Smith, B., Day-Richter, J., Noy, N.F., Musen, M.A.: National Center for Biomedical Ontology: Advancing Biomedicine through Structured Organization of Scientific Knowledge. OMICS A Journal of Integrative Biology 10(2), 185–198 (2006)

[12] Fielding, R.T., Taylor, R.N.: Principled Design of the Modern Web Architecture. ACM Transactions on Internet Technology 2(2), 115–150 (2002)

[13] Xuan, W., Dai, M., Mirel, B., Athey, B., Watson, S., Meng, F.: Interactive Medline Search Engine Utilizing Biomedical Concepts and Data Integration. In: BioLINK SIG: Linking Literature, Information and Knowledge for Biology, Vienna, Austria (July 2007)

[14] Aronson, A.R.: Effective mapping of biomedical text to the UMLS Metathesaurus: the MetaMap program. In: American Medical Informatics Association Annual Symposium, AMIA 2001, Washington DC, USA, pp. 17–21 (2001)

Bio2RDF : A Semantic Web Atlas of Post Genomic Knowledge about Human and Mouse

François Belleau, Nicole Tourigny, Benjamin Good, and Jean Morissette

Centre de Recherche du CHUL, Université Laval
Département d'informatique et de génie logiciel, Université Laval
Bioinformatics Graduate Program, iCAPTURE Centre for Heart and Lung Research,
University of British Columbia

Abstract. The Bio2RDF project uses a data integration approach based on semantic web rules to answer a broad question: What is known about the mouse and human genomes? Using its rdfizing services, a semantic mashup of 65 million triples was built from 30 public bioinformatics data providers: GO, NCBI, UniProt, KEGG, PDB and many others. The average link-rank (ALR) of a node is 4.7 which means that a usual topic is connected to 4.7 other topics by direct or reverse links within the warehouse. A knowledge map of the graph and descriptive statistics about its content are presented. A downloadable version of the Bio2RDF Atlas graph in N3 format is available at http://bio2rdf.org/download.

1 Introduction

According to Davidson [1] the objective of data integration is to make data distributed over a number of distinct, heterogeneous databases accessible via a single interface. Such data integration has been identified as a vital task in the life science domain for more than 20 years and now we are beginning to see promising approaches being delivered; the semantic web based on the RDF model is one of them. RDF (Resource Description Framework)[1] is a metadata model proposed as a standard, by the W3C, to build the emerging semantic web. The Bio2RDF project's main goal is to provide a data integration service to help biologists understand the mechanisms of life and efficiently exploit the vast amount of publicly available data over the web. Applying the semantic web *linked data*[2] approach is its strategy. The DBPedia [2] project uses a map to represent relations between its many linked data sources. Now, if we were to draw a map of the existing relations between linked data from bioinformatics database providers, what would it look like? Could we measure the amount of post genomic knowledge available related to a mouse or human genome sequence? Could it help answer the *what is known* question?

For centuries, maps were essential to give a global representation of a complex and vast reality: our little planet. The explosion of post-genomic knowledge is a

---

[1] http://www.w3.org/RDF/
[2] http://www.w3.org/DesignIssues/LinkedData.html

A. Bairoch, S. Cohen-Boulakia, and C. Froidevaux (Eds.): DILS 2008, LNBI 5109, pp. 153–160, 2008.

consequence of the complete genome sequence being made available after years of DNA sequencing. This scientific accomplishment was possible because genetic maps built of markers existed years before the sequencing began. The 1041 micro-satellite markers of the 1993-1994 Genethon linkage map [3], positioned with odds greater than 1000:1 and distributed over the 23 chromosomes greatly helped the building of the other maps, the hybrid radiation map and the physical one. All these maps were then used to finish the complete genome sequencing project. Without the genetic framework map, the big picture would have been harder or impossible to assemble. What will the framework map be for the successful assembly of the atlas of post-genomic knowledge? Just as we did with the genetic maps, we need a protocol for building the knowledge map and some metrics to describe it.

The bioinformatics community has been actively involved in semantic web development since the first Semantic Web for Life Sciences Workshop[3]. Projects that were first presented, for instance YeastHub [4] and the RDF version of UniProt[4], have shown the way for many other research projects based on RDF, Bio2RDF [5] being one of them. An active community of researchers, members of the HCLS[5], has built a semantic web demo using the Virtuoso server[6]. This project goal is similar to that of Bio2RDF: to make available bioinformatics data in RDF format. Despite these early efforts, there are still not enough mature applications to help scientists because of the semantic web's slow "creep" into the bioinformatics domain [6] and the lack of a naming convention to identify topics [7]. It is still a creepy nation.

In this paper we present the main results of the Bio2RDF project in addressing the problem of data integration in the post genomic era. We will describe the method that was developed to build an RDF graph of 65 million triples, 8 million topics from 30 different public databases. In the results section, we will present statistics used to draw a knowledge map and we will explain why this unified graph contains more knowledge than its parts taken individually. We conclude by proposing future research directions.

2 Semantic Web Ranking

Measuring the amount of links in the WWW and sorting search results according to ranking algorithms have been done successfully with Google's PageRank [8]. Applying the same kind of approach to semantic data was done by the Aleman-Meza group [9]; we adapted his method to define computable statistics quantifying the knowledge within an RDF graph: Openness Ratio, Average Link Rank and Semantic Weight.

Openness Ratio. In a relational database a key not defined within the database cannot be referenced; it is the closed world assumption. According to the open world assumption[7], a corner stone of semantic web architecture, we can link objects together even if we are not sure that destination location exists: RDF graphs can reference

---

[3] http://www.w3.org/2004/07/swls-ws.html

[4] http://dev.isb-sib.ch/projects/uniprot-rdf/

[5] http://www.w3.org/2001/sw/hcls/

[6] http://virtuoso.openlinksw.com/wiki/main/

[7] http://en.wikipedia.org/wiki/Open_world_assumption

undefined URIs. How can we describe the openness cf an RDF graph? How can we measure the fact that this database contains many URIs external to its own domain? We need a connectivity measure at the domain level to measure this type of interrelation. A collection of topics (html documents converted to RDF) is a graph whose subject (Subj) URIs all belong to the same namespace: *geneid* for example. We define the Openness Ratio (OR) of a graph as the proportion of destination URIs (objects of type resource |ObjRes|) which are not also defined within the graph. OR is defined by a normalized difference of cardinality of these two sets and varies between 0 and 1:

$$\mathrm{OR} \;=\; \frac{|ObjRes| - |ObjRes \cap Subj|}{|ObjRes|} \tag{1}$$

A closed database like the MeSH medical vocabulary has an OR of 0 because it does not contain any reference to external databases, all ObjRes are also Subj. At the opposite, OR of Entrez Gene is 1, meaning that all topics contain only URI's outside the *geneid* namespace which make this domain highly open to the external world.

Average Link Rank. How can we now describe the connectivity at the topic level? We define the link rank (LR) as the sum of inbound links (IL) and outbound links (OL) for a specific topic. OL is the number of objects of type resource in the topic graph. IL is the number of topics pointing to it. OL is defined by the topic's object list URIs; IL can only be evaluated by counting topics referring to it and, to do so, the whole global graph needs to be crawled. Before we calculate the Average Link Rank (ALR) for all topics of a graph, a correction needs to by applied to OL; the (1-OR) factor removes all OL not defined within the graph. Finally, to compute ALR, we compute IL and OL individually for each subject, we apply the openness correction and we divide by the number of topics (|Subj|).

$$ALR \;=\; \frac{\sum_{i=1}^{|Subj|} IL_i + (1 - \mathrm{OR}) \sum_{i=1}^{|Subj|} OL_i}{|Subj|} \tag{2}$$

OL of *geneid:15275* topic is 234, but it is impossible to compute IL before the mashup is constructed; in the actual Atlas version its IL is 10. In bioinformatics, IL is analogous to the reverse link defined by KEGG's LinkDB [10]. Obtaining the list of reverse links to a specific document is made easy when RDF data is loaded into a triplestore.

Semantic weight. Heterozygosity refers to the fraction of individuals in a population that are heterozygous for a particular locus. This ratio is critical in linkage analysis, the higher it is, the higher the genetic information content needed to help order a set of genetic markers. It is a measure of genetic information. Could we have a similar indicator to measure the potential of knowledge inside a graph? Our main hypothesis is that knowledge is present in the edges of the RDF graph. We express this as the semantic weight of the graph (SW), the number of relations between two topics in the graph.

$$SW = \frac{|Subj| \times ALR}{2} \tag{3}$$

According to this equation, SW is proportional to ALR. To augment the value of SW we must add new topics, and maximize ALR. To do so, we must minimize OR.

3 Materials and Methods

Selection of the 30 data sources used to build the mashup was done according to three criteria: 1) the database needs to be public, no restriction about data usage should be imposed; 2) the database namespace should be popular among the scientific community, for example *go, pubmed, uniprot* and *omim*; 3) the database should belong to one of these data categories: model organism domain, gene or protein annotation and homology, literature, pathway or chemical. The complete list of data sources with links to reference and copyright is available online in Freebase[8].

The Bio2RDF project uses the Sesame[9] version 1.2.5 triplestore with a MySQL backend repository. Only the relational database implementation could scale to accomodate 65 million triples. Even with current RDF technology, loading millions of triples in the same triplestore is not simple. Once the 30 repositories were individually loaded in separate repositories, their content was moved into the final global graph. We have developed an N3 format extraction program that uses the MySQL backend to extract triples; without this hack, merging all the graphs together would not have been possible. SQL queries are used to compute IL and OL statistics for each topic and to obtain the cardinality of the needed sets.

4 Results

All the high quality graphics in this section were produced using IBM's Many Eyes[10] free visualization service; they can be viewed interactively[11] and the corresponding data set can be downloaded[12]. Once each data source was stored in its own distinct graph, we count the triples whose object referred to external namespaces. The 30 rdfized data sources access 225 different namespaces. Figure 1 illustrates the connectivity between them, each namespace being a node. The MGI and HGNC official gene list resources are highlighted; they are naturally located in the center of the graph because of high OR.

Figure 2 illustrates the same namespace to namespace relationship from a different perspective; it shows the proportion of triples linked to an external topic from each of the 30 data sources. Each color band represents an origin namespace. Bigger circles belong to popular namespaces, the ones frequently used for annotation: *genbank,*

---

[8] http://bio2rdf.org/atlas/sources

[9] http://www.openrdf.org/

[10] http://services.alphaworks.ibm.com/manyeyes/home

[11] http://bio2rdf.org/atlas/map

[12] http://bio2rdf.org/atlas/statistics

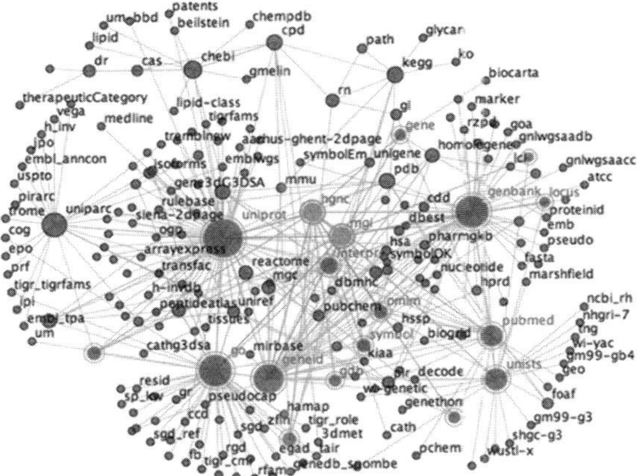

Fig. 1. Bio2RDF map of post-genomic knowledge about human and mouse

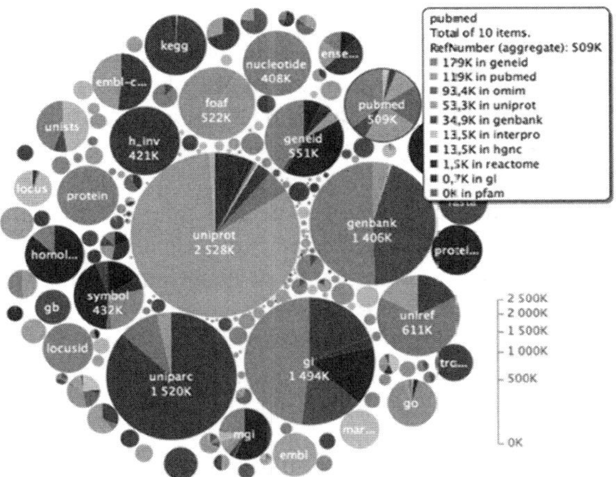

Fig. 2. Bubble chart of InboundLink connectivity between the 225 namespaces

uniprot, etc. We can see that the *pubmed* namespace is used within 10 different domains.

We can also analyze the structure of each RDF topic by namespace at the graph level. Figure 3 illustrates the average size of the RDF graph associated with each subject (topic) in terms of number of triples by subject, literal length by subject and the global OR of the namespace. Those three variables were chosen because they describe three different aspects of the semantics of the graph: 1) the size of the document (triples by subject); 2) the quantity of text (total literal length by subject; 3) the connectivity between namespaces (OR).

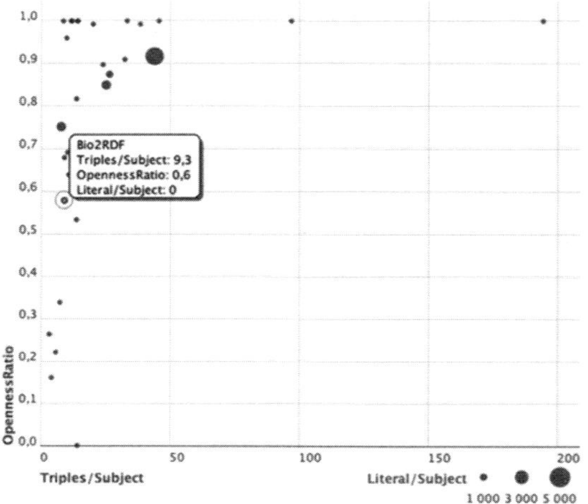

Fig. 3. Descriptive statistics for each of the 30 rdfized data sources

The distribution of the dots in this scatterplot should be familiar to someone working with bioinformatics databases. With an OR of 0.58, the Bio2RDF Atlas has a lower OR than the majority of domains. The biggest dots correspond to *omim, genbank* and *biocarta,* three very literate resources. The domain with the biggest topics are *ec* and *prosite.*

The whole graph is made available to the life science community. The uncompressed file of 9 gigabytes in N3 format can be downloaded[13]. Data is made available in respect to the original license agreement of each data source provider. Bio2RDF original work is published according to a Creative Commons license.

5 Discussion

In the introduction, we made the statement that there is more knowledge in the whole mashup than in all the graphs taken separately. We can now say that the knowledge available in the global graph is higher with 65 million triples merged together. The final Atlas graph OR is 0.58 and its ALR is 4.7. This means that the 8 million topics are connected together by 18.8 millions ((8 x 4.7)/2) relations but also that 58% of all destination URIs are not defined within the Atlas.

Back to our initial hypothesis about knowledge being in the graph, we now propose an explanation. The graph is composed of RDF documents interrelated by URIs; when the graph contains many links to external data sources, OR is near 1. When an RDF warehouse built from different databases is constructed, if its OR is near 0 it means that we have created a closed world database. By merging interrelated graphs together into a mashup, we decrease the OR of the global graph and we increase the ALR of each individual document by adding new IL to it. The average OR of the 30 individual graphs is 0.77 and the OR of the atlas global graph is 0.58; a knowledge

---

[13] http://bio2rdf.org/download

gain of 0.19 obtained by the mashup process. The more connected the graph of linked topics, the more relations are present from which more knowledge can be inferred and, eventually, queried. This is what happens when we build a combined graph of many database sources from independent providers, this is why we have built the Bio2RDF Atlas.

'Critical mass', 'symbiosis' and 'recombinant data' are scientific terms used to describe complex phenomena in nature, they are now used to describe what is observed in the semantic web. Tim Berners-Lee[14] considers that the semantic web in life science will gain a critical mass that will eventually boost the potential of semantic web and linked data technology. Maybe the critical mass could be measured by some metrics like the OR and ALR. Eric Miller[15] considers mashup of RDF to be recombinant data; he meant that RDF data naturally reorganizes, like alleles in a chromosome after a recombination event, without losing functionality. We believe that this auto-organization of information needs URIs to be normalized for the data to recombine, like Bio2RDF does. From this point of view, we consider URIs to be knowledge markers, a place where recombining events occur. Carole Goble [11] suggested that there is a mutual benefit, a symbiotic relation in fact, between e-Science and the semantic web. On one hand, the Bio2RDF mashup could not have been done without the availability of new semantic web tools like Sesame. On the other hand, the life science data domain offers a rich experimental environment to build semantic applications because well-annotated, highly connected (OR > 0.8) data is available. The Bio2RDF project demonstrates what kind of application can emerge from this kind of symbiotic system.

Querying the 65 million triple graph to answer complex questions is the next step. The goal of future work by the Bio2RDF team will be to offer a SPARQL endpoint for the Atlas graph and a full text search service. By making the graph available for download, we invite the triplestore developers to try their software with this data. The current Bio2RDF graph contains a fraction of the linked data available in the genomic domain; many billions of triples are waiting for a semantic technology to query them with finite resolution time.

6 Conclusion

According to Tim Berners-Lee, the inventor of HTML and an enthusiastic proponent of the semantic web, life science may provide a killer application for demonstrating the usefulness of the semantic web approach. We certainly agree with that affirmation and we believe that such a convincing application could arise from the Bio2RDF project. How fast will the scientific community adapt to this new paradigm of sharing knowledge using semantic web technologies? How can the existing knowledge be intelligently used to help researchers make discoveries? In the 17th century, the invention of the telescope opened the door to modern astronomy and the vastness of the universe became real; the microscope was the starting point for microbiology. Those devices in the hands of scientists have forged our understanding

---

[14] http://www.bio-itworld.com/newsitems/2005/05/05-19-05-news-Berners-Lee
[15] http://www.bio-itworld.com/issues/2006/june/mashups/

of nature. How to navigate, sort, discover, zoom in and zoom out in our 21st century universe of linked data? We need appropriate tools to magnify our intelligence, to focus our research, to create links between new documents, between new ideas that are part of the living web of knowledge shared by the social network of science. Now that a sky of data is available, now that we need to find a lost page in the gigantic web, new devices are needed to assist scientists. We have started to map the knowledge space of biology, we have a first impression of what the bioinformatics nation looks like, the time has come to explore it, the time has come to build the *knowledgescope*.

References

[1] Davidson, S.B., Overton, C., Buneman, P.: Challenges in integrating biological data sources. J. Comput Biol. 2, 557–572 (1995)
[2] Auer, S., Bizer, C., Lehmann, J., Kobilarov, G., Cyganiak, R., Ives, Z.: PDF DocumentDBpedia: A Nucleus for a Web of Open Data. In: Aberer, K., Choi, K.-S., Noy, N., Allemang, D., Lee, K.-I., Nixon, L., Golbeck, J., Mika, P., Maynard, D., Mizoguchi, R., Schreiber, G., Cudré-Mauroux, P. (eds.) ISWC 2007 + ASWC 2007. LNCS, vol. 4825. Springer, Heidelberg (2007)
[3] Gyapay, G., Morissette, J., Vignal, A., Dib, C., Fizames, C., Millasseau, P., Marc, S., Bernardi, G., Lathrop, M., Weissenbach, J.: The 1993-94 Généthon human genetic linkage map. Nat. Genet. 7, 246–339 (1994)
[4] Cheung, K., Yip, K.Y., Smith, A., Deknikker, R., Masiar, A., Gerstein, M.: YeastHub: a semantic web use case for integrating data in the life sciences domain. Bioinformatics 21(suppl. 1), i85–i96 (2005)
[5] Belleau, F., Nolin, M.-A., Tourigny, N., Rigault, P., Morissette, J.: Bio2RDF: Towards A Mashup To Build Bioinformatics Knowledge System. Journal of Biomedical Informatics (in press, 2008) doi:10.1016/j.jbi.2008.03.004
[6] Good, B.M., Wilkinson, M.D.: The Life Sciences Semantic Web is full of creeps! Brief Bioinform 7, 275–286 (2006)
[7] Goble, C., Stevens, R.: State of the nation in data integration for bioinformatics. Journal of Biomedical Informatics (online, 2008) doi:10.1016/j.jbi.2008.01.008
[8] Brin, S., Page, L.: The anatomy of a large-scale hypertextual Web search engine. Computer Networks and ISDN Systems 30(1-7), 107–117 (1998)
[9] Aleman-Meza, B., Halaschek-Wiener, C., Arpinar, I.-B., Ramakrishnan, C., Sheth, A.P.: Ranking Complex Relationships on the Semantic Web. IEEE Internet Computing 9(3), 37–44 (2005)
[10] Fujibuchi, W., Goto, S., Migimatsu, H., Uchiyama, I., Ogiwara, A., Akiyama, Y., Kanehisa, M.: DBGET/LinkDB: an integrated database retrieval system. In: Pac. Symp. Biocomput., pp. 683–694 (1998)
[11] Goble, C., Corcho, O., Alper, P., De Roure, D.: e-Science and the Semantic Web. In: Balcázar, J.L., Long, P.M., Stephan, F. (eds.) ALT 2006. LNCS (LNAI), vol. 4264, p. 12. Springer, Heidelberg (2006)

OMIE: Ontology Mapping within an Interactive and Extensible Environment

Amel Bouzeghoub[1], Abdeltif Elbyed[1,2], and Fariza Tahi[2]

[1] Institut Telecom SudParis, Evry, France
`{Amel.Bouzeghoub,Abdeltif.Elbyed}@it-sudparis.eu`
[2] IBISC FRE CNRS 3190, Université d'Evry-Val d'Essonne, Genopole, France
`{abdeltif.albyed,fariza.tahi}@ibisc.univ-evry.fr`

Abstract. The increasing number of on-line accessible biological data sources has involved a growth of the number and the size of ontologies. This makes it increasingly valuable to map ontologies each other to determine which of their concepts are semantically related. Nowadays, developed tools are often semi-automatic and require the help of experts. Determining semantic relations between concepts is a difficult task, and the problem is still open. Several methods have been proposed in the literature. Existing tools for mapping concepts usually combine several methods, called matchers. In this paper, we propose a tool called OMIE (Ontology Mapping within an Interactive and Extensible environment) which uses and combines several matchers. OMIE is extensible, i.e. matchers could be added or inhibited, and is interactive, i.e. experts could validate or invalidate mappings as well as choose between mapping specific concepts or mapping the entire ontologies.

Keywords: biomedical and life science ontologies, ontology mapping, matchers, similarity measures, semantic web, multi-agent systems.

1 Introduction

Ontologies are increasingly used and become important in several areas, including life sciences. They are used as basis for interoperability between systems and for data integration, by providing a common terminology over a domain. In life sciences, several ontologies have been developed, in order to cover specific domains, e.g. Gene Ontology (GO), Foundational Model of Anatomy (FMA), Adult Mouse Anatomy (MA), etc. Most of these ontologies are accessible from the Open Biomedical Ontologies (OBO) website. These ontologies are mostly complementary but contain important overlapping.

Nowadays, there is an increase of bioinformatics projects which need to use several ontologies related to complementary domains of life sciences. The SAPHIR project (a Systems Approach for PHysiological Integration of Renal, cardiac and respiratory functions) [1], is one of them. In order to use the needed ontologies in an integrated way, "bridges", i.e. mappings, between the ontologies must be built. Mapping two ontologies O1 and O2 means defining semantic relations between

A. Bairoch, S. Cohen-Boulakia, and C. Froidevaux (Eds.): DILS 2008, LNBI 5109, pp. 161–168, 2008.

concepts of O1 and concepts of O2. Mappings are often established manually by experts, but because of the increase number of life sciences ontologies and the increase of their size, there is a need of automatic (or semi-automatic) mappings. To infer semantic links between two concepts or terms, several methods, called matchers, have been proposed, but none of them is able to insure good results alone. Thus, the often adopted solution is to combine several matchers. In this paper, we propose an interactive, extensible and distributed tool for automatic and semi-automatic mapping ontologies. The prototype, called OMIE (Ontology Mapping within and Interactive and Extensible Environment), uses and combines several matchers. The matchers generate mapping hypotheses which are then filtered by series of filters. OMIE is extensible, i.e. matchers could be added or inhibited, and is interactive, i.e. experts could validate or invalidate mappings and could choose specific concepts to map, or map the entire ontologies. Furthermore, OMIE offers to users the possibility to give their feedback on mapping results. Theses feedbacks are then used to improve the mappings. The prototype is a multi-agent based- system and is therefore completely distributed.

The paper is structured as follows: in the next section we relate briefly works done on ontology mapping in life sciences, then we describe our system OMIE. The results are illustrated in section four with two biomedical ontologies, namely, Medical Subject Headings (MeSH) and the Mouse Anatomy (MA). The paper ends with conclusion and remarks on further works.

2   Related Works

Very few ontology mapping tools are used in the context of biology. PROMPT [2] is perhaps the most known, since it is a plugin-in of Protégé-2000 [3], the most popular ontology editor in the community of biologists. In [4], an evaluation of PROMPT in Protégé-2000 and Chimaera [5] was done on biological ontologies. Two ontologies were considered: Gene Ontology (GO) and Signal-Ontology (SO). The conclusion was that both tools are helpful, but not completely satisfactory, either in their use or in the results they give.

Mapping results depend considerably on methods used for comparing concepts and inferring the semantic relations between them. In the literature, we can find several proposals of methods for mapping and/or aligning biological ontologies. Most of them are systematic and devoted in particular ontologies: in [6] and in [7], they were interested by aligning FMA and GALEN, two representations of anatomy; in [8], they were interested by aligning mouse and human anatomies, the Adult Mouse Anatomical Dictionary and the NCI Thesaurus (human anatomy). Others are more general. In [9], a framework called SAMBO for mapping and merging biomedical ontologies is presented and compared to PROMPT and also to FOAM, another well-known ontology mapping tool [10]. Finally, an algorithm for matching life sciences ontologies and based on instances is proposed in [11]. The authors have suggested the use of data contained in the data source Ensembl in order to find potential mappings between concepts. Their method was tested on GO and OMIM ontologies.

3   OMIE Description

3.1   Functionalities

The functionalities offered by OMIE can be divided into three main categories:

- **Query mode:** a user may need to map his/her local ontology with a chosen
  target ontology semantically closed or to map only a part of his/her local
  ontology with a target ontology. To carry out this goal we propose two ways
  of mapping: (i) 'concept mapping' to map merely selected concepts and (ii)
  'global mapping' to map the entire ontology. At the end of the mapping
  process, the user has the possibility to give his/her satisfactory level of the
  obtained results (feedbacks).
- **Admin mode:** the administrator may propose specific matchers or filters.
  According to the application domain, he/she may fit, deactivate or refine all
  application variables (i.e., thresholds, matcher confidences, etc.). To provide
  more flexibility to our approach, we propose different matchers and filters at
  the beginning of the mapping process. The administrator may change this
  standard configuration.
- **Expert mode:** experts may interact with the system to validate or invalidate the generated mappings. During the validation process, they are asked
  (by the system) for an 'expertise level', information used in the mapping process (see below). They have also the possibility to add mappings by hand.

3.2   The Mapping Process

Given two ontologies O1 and O2, OMIE determines mappings between concepts
of O1 and concepts of O2 using several similarity methods which are of four
types: syntactic, linguistic, structural and semantic. OMIE is a multi-agent-
based system composed of five principal types of agents, each agent having a
specific role (Figure 1):

- Ontology Agents (OA): each ontology agent is associated to one ontology.
  The user interacts with these agents to send his/her mapping query which
  is then handled by MA agents.
- Matcher Agents (MA): each MA agent calculates a similarity value for each
  couple of concepts.
- Hypotheses Generation Agent (HGA): combines the similarity values generated by the different Matcher Agents and generates a mapping hypothesis
  for each couple of concepts.
- Hypotheses Filtering Agent (HFA): its role is to filter the mapping hypotheses sent by HGA agent in order to eliminate the least valid ones.
- Feedback and Validation Agent (FVA): interacts with the user for the validation of the mapping hypotheses.

The system is completely distributed, and all agents work separately.

The mapping process in OMIE is thus composed of four main steps: Similarity
computation, candidate mapping generation, candidate mapping filtering and
user validation and feedbacks. Each step is performed by a specialized agent.

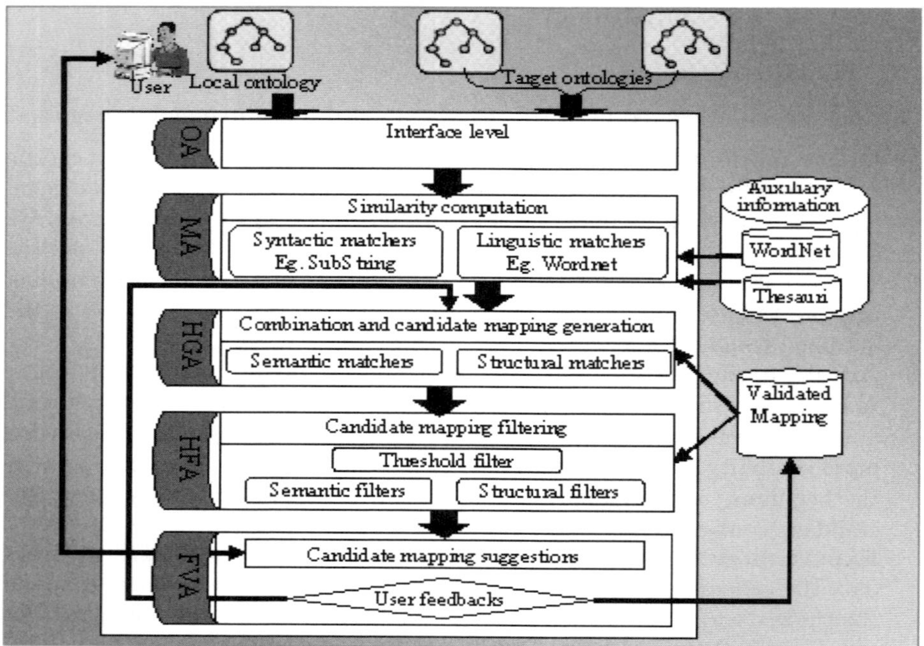

Fig. 1. General architecture of the mapping process in OMIE

Similarity computation: OMIE executes multiple independent matchers (each matcher is implemented by an agent) in order to measure the similarity value between two given concepts. In this step, two types of matchers are considered: linguistic matchers and syntactic matchers. Matcher agents (MA) work at the same time and independently. We have chosen and implemented different existing solutions: WordNet for linguistic comparisons and several functions for syntactic comparisons (String Equality function, Similarity String function, Levenshtein function, Sub-String Distance function and the Hamming Distance function). For each considered couple of concepts, each MA agent returns a similarity value (between 0 and 1). Besides these matchers, experts and administrators may propose other specific matchers to improve the result's quality.

Candidate mapping generation: HGA agent collects all similarity values generated by the different MA agents. It creates a set of possible candidate mappings (called hypotheses). Each hypothesis Hp is a tuple $< c, d, Conf_{Hp}, SV_{Hp} >$ such as: $Conf_{Hp} = \sum_i Conf_{h_i}$ and $SV_{Hp} = \frac{\sum_i Conf_{h_i} SV_{h_i}}{\sum_i Conf_{h_i}}$ and where $Conf_{h_i}$ and SV_{h_i} are respectively the confidence level of the matcher h_i and the similarity value returned by the matcher h_i for the couple of concepts c and d. Structural (topological) and semantic matchers are here used to propose new hypotheses and/or improve similarity values of existing hypotheses. HGA exploits the hierarchical structure and the semantic relations of the ontology, as well as the mappings already established and validated in previous iterations.

Candidate mapping filtering: We have developed several methods to filter hypotheses generated by HGA agent. HFA agent detects and solves incompatibilities between mappings (e.g. when in one hand, two concepts c and d are mapped and in another hand, a father of c is mapped with a child of d (crossed mappings)). Thus, if a mapping hypothesis is incompatible with a validated mapping, it is eliminated; and if two mapping hypotheses are incompatible, the one having the lowest similarity value is eliminated. Besides, we have defined a threshold on the similarity value SV_{Hp} and on the confidence level $Conf_{Hp}$ under which the mapping hypothesis is not considered. Of course, these thresholds could be changed by the user.

Accurate candidate mapping validation and user feedback recovering: In this step, FVA agent interacts with the user in order to propose the mapping hypotheses sent by HFA agent. The user could validate or invalidate each of the mapping hypotheses. When a mapping is validated, it is stored, and then used by HGA and FHA agents in order to improve the mapping process (see above).

An important aspect that completes the interactive notion is the user feedback. The user can express directly his/her satisfactory level on the obtained results. We have associated a "validation level" to each mapping. This validation level depends on the number of times the mapping has been validated by users and on the "expertise level" of the users (given by the users themselves). The level of expertise is a value between 0 and 1. When this level is set to 1 by an expert for a given mapping, this would mean that he is completely sure about the validity of the mapping. Thus, the validation level of a mapping, initially set to zero, increases with an expertise level at each validation by an expert.

4   Results

OMIE is implemented using the multi-agent platform JADE (Java Agent Development Framework). We use OntoBroker system [12] to manage the ontologies. OntoBroker integrates various input formats of ontologies like RDF(S), F-Logic or OWL. The different matchers and similarity methods we developed are implemented with logic rules, which make OMIE easily extensible.

A great number of available biomedical ontologies are in OBO format. Concepts in these ontologies are described with identifiers (e.g., MA_00003, MA_00005, and so on) and label properties (e.g., organ system). This makes the mapping impossible if we consider only identifiers. Contrary to Protégé and PROMPT which browse and map only concept identifiers, OMIE is able to edit and map both identifiers and labels.

We tested OMIE on several biomedical ontologies. We present here the results obtained with two ontologies, namely: MeSH (Medical Subject Headings) ontology [13] and MA (Adult Mouse Anatomy) ontology [14]. These ontologies cover a similar anatomy context and are developed independently. MeSH is a controlled vocabulary produced by the American National Library of Medicine and is used for indexing, cataloguing, and searching for biomedical and health-related information and documents. It consists of sets of terms or descriptors in a hierarchical structure and contains more than 1400 concepts. MA organizes anatomical structures for the postnatal mouse spatially and functionally, using 'is a' and 'part of' relationships. The ontology is used to describe expression data for adult mouse and phenotype data pertinent to anatomy in standardized ways. MA ontology contains more than 2400 anatomical concepts. For our experimentation we focused on three categories developed by both ontologies, namely, nose (with 15 concepts in MeSH and 18 concepts in MA), ear (39 concepts in MeSH and 77 concepts in MA), and eye (45 concepts in MeSH and 112 concepts in MA).

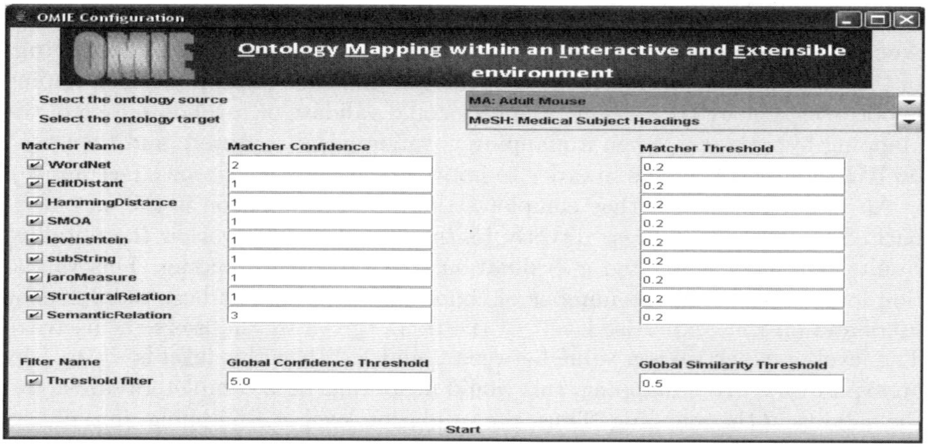

Fig. 2. OMIE configuration interface

Figure 2 shows the system configuration used in our tests: selected matchers, confidence level and similarity value threshold associated to each matcher, and global thresholds (similarity value and confidence level) used in the filtering step. All our test evaluations are based on the metrics of recall and precision calculated considering mappings generated by the system and mappings identified manually by domain expert. In our tests, the domain expert provides 9 mappings between nose concepts, 27 mappings between ear concepts and 27 mappings between eye concepts.

We compared the results obtained by OMIE on MeSH and MA ontologies with the ones obtained by Lambrix and Tan in [9], where they evaluated their ontology mapping tool SAMBO with two well known and available tools: PROMPT [2] and FOAM [10]. Figure 3 shows the recall and the precision obtained by each of the four systems (PROMPT, FOAM, SAMBO and OMIE). We can see that the precision of OMIE is higher than the ones of PROMPT and FOAM and is equal to the precision of SAMBO, and that the recall of OMIE is the higher one. For example, OMIE succeses to generate all mappings provided (manually) by the domain expert for the 'eye' concepts.

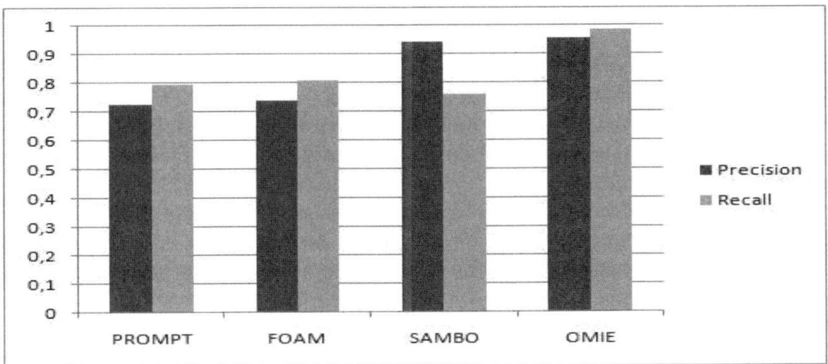

Fig. 3. Comparison of mapping results obtained by OMIE, PROMPT, FOAM and SAMBO on MA and MeSH ontologies

5   Conclusion

We propose in this paper a system, called OMIE, for mapping ontologies, in particular biological ontologies. This system, implemented under a multi-agent approach, is distributed and more importantly is interactive and extensible. Thanks to an ergonomic interface, users can choose between mapping specific concepts or all concepts of two selected ontologies. They can also validate or invalidate mappings. Administrators can choose, add or inhibit matchers, i.e. similarity measure procedures. One of the characteristics of OMIE is the ability to use dynamically validated mappings in order to improve the mapping process in generating mappings as well as in detecting wrong mappings. We use several similarity measure methods of different types: syntactic, linguistic, topological and semantic. These methods are combined in order to increase the chance of generating efficiently the mappings. Other methods could be used. Thus, we are working on developing other matchers. An important one is an instance-based matcher: two concepts could be mapped if they are associated to sets of data (sources) or documents which are similar. Like in [15] we propose to use ontology instances to enrich the ontology by creating new semantic relations between ontology concepts, which could be used by HGA agent (semantic matchers) as well as FVA agent (semantic filters). Finally, we plan to use automatic learning methods to make more reliable the confidence levels and the similarity value thresholds associated to each matcher as well as the thresholds of global similarity value and of global confidence level used in the filtering step.

Acknowledgement. This work is supported in part by the National Office of Research (ANR) Biosys ("SAPHIR" project) and by the regional council of Essonne" ("POPS" project in the competitive cluster "System@tic").

References

1. Thomas, S., Abdulhay, E., Baconnier, P., Fontecave, J., Francoise, J., Guillaud, F., Hannaert, P., Hernandez, A., Rolle, V.L., Maziere, P., Tahi, F., Zehraoui, F.: Saphir - a multi-scale, multi-resolution modeling environment targeting blood pressure regulation and fluid homeostasis. In: Conf. Proc. IEEE Eng. Med. Biol. Society, pp. 6649–6652 (2007)
2. Noy, N.F., Musen, M.A.: PROMPT: Algorithm and tool for automated ontology merging and alignment. In: AAAI/IAAI, pp. 450–455 (2000)
3. Knublauch, H., Fergerson, R.W., Noy, N.F., Musen, M.A.: The protégé owl plugin: An open development environment for semantic web applications. In: International Semantic Web Conference, pp. 229–243 (2004)
4. Lambrix, P., Edberg, A.: Evaluation of ontology tools in bioinformatics. In: Pacific Symposium on Biocomputing, pp. 529–600 (2003)
5. McGuinness, D.L., Fikes, R., Rice, J., Wilder, S.: The chimaera ontology environment. In: AAAI/IAAI, pp. 1123–1124 (2000)
6. Zhang, S., Bodenreider, O.: Aligning representations of anatomy using lexical and structural methods. In: AMIA Symposium Proceedings, pp. 753–757 (2003)
7. Mork, P., Bernstein, P.: Adapting a generic match algorithm to align ontologies of human anatomy. In: 20th International Conf. on Data Engineering. IEEE, Los Alamitos (2004)
8. Bodenreider, O., Hayamizu, T., Ringwald, M., Coronado, S.D., Zhang, S.: Of mice and men: Aligning mouse and human anatomies. In: AMIA Annu. Symp. Proc., pp. 61–65 (2005)
9. Lambrix, P., Tan, H.: SAMBO–a system for aligning and merging biomedical ontologies. Web Semantics: Science, Services and Agents on the World Wide Web 4(3), 196–206 (2006)
10. Ehrig, M., Sure, Y.: FOAM - framework for ontology alignment and mapping - results of the ontology alignment evaluation initiative. In: Proc. of the Workshop on Integrating Ontologies, vol. 156, pp. 72–76 (2005)
11. Kirsten, T., Thor, A., Rahm, E.: Instance-based matching of large life science ontologies. In: Cohen-Boulakia, S., Tannen, V. (eds.) DILS 2007. LNCS (LNBI), vol. 4544, pp. 172–187. Springer, Heidelberg (2007)
12. Ontoprise: OntoBroker user guide, http://www.ontoprise.de
13. National Library of Medicine: Medical Subject Headings (MeSH), http://www.nlm.nih.gov/mesh
14. Hayamizu, T., Mangan, M., Corradi, J., Kadin, J., Ringwald, M.: The adult mouse anatomical dictionary: a tool for annotating and integrating data. Genome Biology 6(3) (2005)
15. Bouzeghoub, A., Elbyed, A.: Ontology mapping for web-based educational systems interoperability. IBIS 1(1), 73–84 (2006)

Chemical Knowledge for the Semantic Web

Mykola Konyk[1], Alexander De Leon[1], and Michel Dumontier[1,2,3]

[1] School of Computer Science
[2] Department of Biology
[3] Institute of Biochemistry,
Carleton University, 1125 Colonel By Drive,
K1S 5B6, Ottawa, Canada
mkonyk@gmail.com, alexjdl@gmail.com,
michel_dumontier@carleton.ca

Abstract. With over 80 file formats to represent various chemical attributes, the conversion between one format and another is invariably lossy due to informal specifications. In contrast, the use of a formal knowledge representation language such as the Web Ontology Language (OWL) enables precise molecular descriptions that can be reasoned about in a logically valid manner. In this paper, we describe a chemical knowledge representation using OWL. We demonstrate its utility in querying a new drug repository created from PubChem, DrugBank and DBpedia. By leveraging Semantic Web technologies, it becomes possible to integrate chemical information at differing levels of detail and granularity, opening new avenues for life science knowledge discovery.

Keywords: semantic web, knowledge representation, knowledge engineering, ontology, life sciences, question answering, OWL, chemistry, molecule, mashup.

1 Introduction

While powerful web search engines can sift through enormous amounts of biochemical information online, it is still difficult to find compounds having a set of desirable attributes i.e. can form specific derivatives, or are stable at room temperature and have a non-toxic metabolic profile. Although over 80 file formats exist to represent chemical data, none, including the Chemical Markup Language (CML) [1], are capable of encoding arbitrarily knowledge in such a way that the meaning is wholly preserved. Controlled vocabularies have been designed for chemical functional groups (CO [2]) or compounds (ChEBI [3]), but they are generally used for the annotation of chemicals or in navigation of search results. In contrast, Semantic Web ontologies aim to explicitly describe and relate objects using formal, logic-based representations that a machine can understand and process [4]. This will facilitate knowledge representation, integration and question answering in areas of critical importance to the life sciences.

In this paper, we describe a knowledge representation for chemical information using OWL, the Web Ontology Language [5]. OWL facilitates the description of

A. Bairoch, S. Cohen-Boulakia, and C. Froidevaux (Eds.): DILS 2008, LNBI 5109, pp. 169–176, 2008.

complex concepts from simpler ones and can be used for consistency checking and classification [6]. We describe our efforts to integrate DrugBank and PubChem, two popular chemical databases and DBpedia, an RDF version of Wikipedia. Finally, we illustrate the value of using semantic web technologies to seamlessly integrate and query diverse biochemical knowledge in a manner that opens new avenues for knowledge discovery in the life sciences.

2 Methods

2.1 Chemical Knowledge Representation

Upper level ontologies increase interoperability and semantic coherency of domain ontologies by grounding the basic types of domain entities and imposing restrictions on the relationships that these entities may hold. We use the Basic Formal Ontology (BFO) [7] because it offers a simple framework that distinguishes objects, qualities, processes and spatial regions. Our Basic Relation Ontology[1] (BRO) provides object-process, object-quality, parthood, spatial, temporal relations drawn from foundational work [8]. The New Upper Level Ontology[2] (NULO) maps the domain and range values of BRO properties to BFO concepts, and further constraints on relations are specified in NULO-constraints[3]. Reflexive, irreflexive, asymmetric, disjoint roles and role chains have been added to the BRO-OWL11 ontology[4] so as to maximize reasoning capability [9].

An outline of the chemical knowledge representation is illustrated in Fig 1. Briefly, molecules, atoms and rings are types of objects that bear qualities and may be located in spatial regions.

Objects: Molecules, atoms, rings are types of objects that are spatially extended, maximally self-connected and self-contained and bear any number of qualities appropriate to their type.

Qualities: A quality is a categorical property that exists in some object. Qualities have been defined for each kind of object. For instance, a molecule might bear the quality of monoisotopic mass whereas the partial charge is an atom quality. Some quality types may be borne by multiple types of objects (i.e. atoms or molecules may bear a chiral quality). We have identified over 50 types of qualities, largely defined from OpenBabel and PubChem descriptors.

Mereology: A molecule is composed of at least two or more atoms and has zero or more ring parts. Molecules or Rings are related to Atoms by *hasProperPart,* an asymmetric relation. Molecules and rings are related to each other by *hasPart*, a transitive (if a *hasPart* b and b *hasPart* c, then a *hasPart* c) and reflexive (one can have itself as a part) relation. Thus, rings may also be a molecule (i.e. benzene).

---

[1] http://ontology.dumontierlab.com/bro
[2] http://ontology.dumontierlab.com/nulo
[3] http://ontology.dumontierlab.com/nulo-constraints
[4] http://ontology.dumontierlab.com/bro-owl11

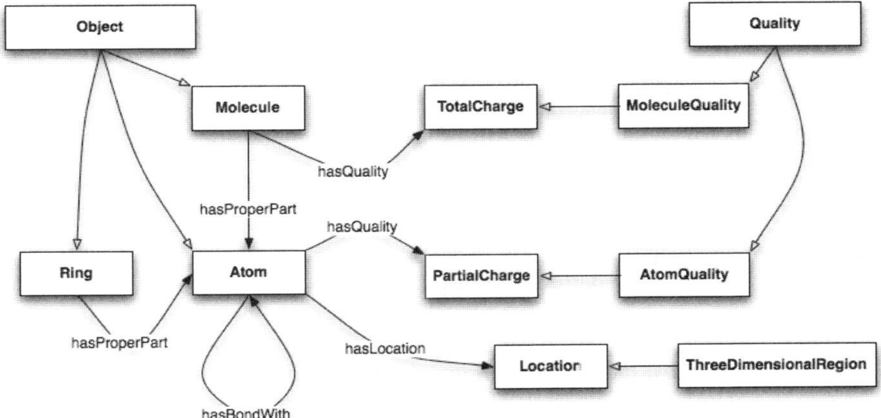

Fig. 1. Overview of major ontological components and their relationships in the chemical knowledge representation

Connectivity: Atoms are connected to each other via symmetric *hasBondWith* object properties. Specification of number of shared of electrons is done via sub-properties (e.g. *hasSingleBondWith, hasAromaticBondWith*).

Stereochemistry: The spatial arrangement of atoms within molecules affects behavior and function. Stereochemical knowledge is reflected at the molecule (the molecule is a ChiralMolecule), atom (the atom is a ChiralAtom) and bonds (*hasWedgeBondWith* and its inverse *hasHashBondWith*) levels.

Location: Physical objects such as molecules or atoms may be spatially located in two or dimensional spatial regions to which specific coordinates may be assigned. The Cartesian coordinates of a three dimensional spatial region are assigned via datatype properties (*coordinateX, coordinateY* and *coordinateZ*). Since atoms are parts of molecules, and the region of space that atoms occupy is part of the region of space that molecules occupy, we can say that an atom *isLocatedIn* molecule [8].

2.2 Open Babel: Chemical File Conversion to OWL

We implemented a plugin for the widely used and freely available Open Babel software suite to convert any of the 80 chemical file formats into an OWL chemical knowledge model. OB provides an application programming interface (API) for reading and writing chemical file formats, accessing information about molecules, atoms, bonds, rings and for computing chemical attributes. Since each file format is different and contains an arbitrary set of information, we compute missing information using Open Babel built-in routines, where possible.

The plugin architecture is highly flexible and allows one to create mappings from ontology classes and their attributes to main classes of the OB data model. The mappings are defined within 7 major sections of an XML based configuration file.

Generate: Specifies how the ontology should be generated. For example, it deals with adding comments or time stamps in the ontology header.

Base: Specifies the namespace of the ontology.

URIs: Specifies which namespaces will be used in the ontology.

Import: Specifies which ontologies should be imported to provide additional information of named entities.

Classes: Contains mapping rules for establishing class type and membership. OWL classes allow the grouping data with similar properties by defining the necessary and sufficient conditions for class membership. For example, one can define the HydrogenAtom class as an Atom (to which all of the atoms present in the OB data model get mapped) that have 1 as their atomic number. More complex mappings may be generated through unions and intersections of restrictions (or combinations of both) and nested conditions.

DataProperties: Contains mapping rules to specify datatype properties. Datatype properties describe binary relations between OWL individuals and RDF literals or XML schema datatypes. For example, the Atom class would be the domain of hold the *atomicNumber* datatype property whereas Location would be the domain of *coordinateX*, *coordinateY* datatype properties.

ObjectProperties: Contains mapping rules to specify object properties. Object properties describe relations between OWL individuals. A domain and range may be specified for each property; hence, we may define the *hasProperPart* object property with *Molecule* as a domain and *Atom* as a range. In this case, *hasProperPart* object property will be created between every single atom individual and a molecule individual (an individual is an instance of an OWL class).

2.3 DrugBank

DrugBank is comprehensive drug knowledge base that is freely available on the web [10]. It combines clinical and chemical information about drug molecules and also provides detailed information about their drug targets. DrugBank contains nearly all drugs that have been approved in North America, Europe and Asia. These have been tagged as approved, experimental, biotech, nutraceutical, illicit and withdrawn drugs.

OWL classes are generated from each DrugBank "drugcard" records using Apache Group's open-source implementation of UIMA[5]. UIMA is a framework to analyze large amount of unstructured information using a workflow of annotators. Each annotator uses information from the original input and/or from previous annotators in the workflow and produces new information that is made available to other annotators further in the workflow. We designed an RDF/XML template to allow UIMA annotators to collaborate in converting DrugBank records into an OWL class. This flexible approach decouples the OWL representation from the software.

Drugs are types of objects represented as OWL classes. By importing the ontology into an existing OWL knowledge base, one can automatically classify instances based

---

[5] http://incubator.apache.org/uima/

on their characteristics. For example, the drug Leuprolide is equivalent to the class of all things that have *pubchemcompoundid* = 3911. On reasoning, we discover that all individuals asserted as instances of this class will inherit the property of having *pubchemcompoundid* = 3911 and that an individual that contains the value 3911 for the data property *pubchemcompoundid* will be inferred as an instance of the class.

2.4 DBpedia Integration

DBpedia makes the encyclopedia-like information from Wikipedia available in RDF. We mapped the Wikipedia link found in some DrugBank records to the corresponding DBpedia entry. The corresponding URI was found by querying DBpedia's SPARQL endpoint for the resource that is the subject of the given Wikipedia page. When adding the DBpedia RDF graph, the record is visible to the ontology as an OWL individual. To strengthen the relationship between the DBpedia instance and the corresponding drug class from the Drugbank ontology, we assert that the class is equivalent to the set containing the DBpedia instance. This is expressed in OWL using *enumerations* (owl:oneOf).

3 Results

We created an example OWL knowledge base[6] from some of the i) 4422 UIMA-generated OWL ontologies from DrugBank records with PubChem identifiers, ii) Open Babel plugin generated OWL ontologies from PubChem SDF records and iii) script generated OWL import documents for DBpedia URIs from DrugBank Wikipedia links. We will demonstrate querying this knowledge base using the simple Manchester OWL syntax [11].

Use Case 1: Querying Substructures, Functional Groups and Compounds
An important aspect of chemical synthesis, pharmaceutical design and lead optimization involves searching chemical databases for compounds having certain kinds of substructures. Our knowledge model provides the means to define and search for substructures. As an example, let us search our knowledge base the -OH substructure.

DLQuery: *OxygenAtom that hasSingleBondWith some HydrogenAtom*

Such queries can be captured in an ontology of functional groups. A functional group describes the semantics of chemical reactivity in terms of atoms and their connectivity, and exhibits characteristic chemical behavior when present in a compound. In our ontology of major functional groups found in organic compounds[7], the organic alcohol group is defined as R-OH, where R is any alkyl or aryl carbon. Importing the functional group ontology into the chemical knowledge base enables the reasoner to automatically discover which atoms are part of known substructures, and we can query accordingly:

---

[6] http://ontology.dumontierlab.com/ckb-dils2008
[7] http://ontology.dumontierlab.com/organic-functional-group-complex

DLQuery: *Molecule that hasPart some AlcoholGroup*

An ontology of organic compounds[8] provides the necessary and sufficient conditions to automatically classify molecules based on the presence of functional groups. Hence, this ontology allows us to refer to the encapsulated concept in future queries:

DLQuery: *Alcohol*

In this way, once a substructure or functional group is defined, it can be captured as an ontology concept and published on the semantic web for sharing and reuse.

Use Case 2: Simultaneous Querying of Chemical Qualities and Substructures
A chemical knowledge base generated from the Open Babel OWL plugin will have structural information and a wide variety of descriptors, including identifiers. To ask about the set of descriptors for leuprolide using the PubChem identifier:

DLQuery: *isQualityOf some (Molecule and pubchemcompoundid value 3911)*

Answers to this query involve inferences drawn from i) the domain value of *isQualityOf* and ii) the *hasQuality* inverse property. First, PubChem descriptors are inferred to be qualities of an object due to fact that a Quality is the domain of the *isQualityOf*. Second, the knowledge base contains *hasQuality* assertions between molecules, atoms and rings and since the inverse of *isQualityOf* is *hasQuality*, it is possible to answer this query. Qualities including total charge, heat of formation, molecular mass, among others in the example knowledge base.

Use Case 3: Query over PubChem, DrugBank and DBpedia
Taken together, we can pose a fairly sophisticated query across our expressive ontologies and the three resources to ask about biotech drugs (DrugBank) that have an alcohol moiety (PubChem) and are eliminated within an hour (DBpedia):

DLQuery: *Alcohol and BiotechDrug and eliminationHalfLife value "Hour"*

4   Discussion

Knowledge representation. Representing chemical knowledge using an expressive formal language like OWL enables new opportunities for data integration and classification that are not possible with XML or RDF (on their own). Here, we take a step forward towards a more *realist* representation with respect to how molecules are composed, the qualities they bear, and the spatial locations they occupy. Having a regular and coherent representation rooted in reality should facilitate the classification of a feature and how it will be added to our knowledge.

While our approach is guided by the Basic Formal Ontology, it is insufficient in several respects. The first is that the BFO would like to define types, and ensure that instantiated types are those that really do exist (and that we can point to). Unfortunately, several problems arise. *First*, OWL is inadequate to specify the full

---

[8] http://ontology.dumontierlab.com/organic-compound-complex

molecular structure at the class level. This is because OWL class descriptions may not contain cycles. To overcome this limitation, we define classes in which only a single instance, containing the structural description, is the member. In this way, every instance of that class will inherit the properties of the equivalent instance. However, this approach has a serious consequence: that all instances of that class are equivalent to that single instance and therefore not differ. Thus, it will never be possible to have a collection of instances. So the solution is mostly to integrate information, rather than to have a realistic representation. As such, the solution is unsatisfying. However, recent work [12] to represent structured objects in OWL should prove adequate in this regard, and provide the means by which we can describe full molecular structure at the class level. *Second*, the integration of RDF-data from DBpedia forces an instance-level representation. This is because for a proper class description, triples must be converted into class restrictions, which are syntactically different. This poses a major problem that we overcome in the same manner as was used for molecule structure, but remains restricted to data integration, rather than realistic representation. Moreover, there exists an exceptional challenge in interpreting DBpedia data properties. Some have cryptic single letter names (i.e. "c" or "r") for which no definition is provided. Longer term goals include creating an OWL mapping to DBpedia types identified with molecule records.

Data integration: While data integration is trivially satisfiable when using the same URIs, it is also possible to integrate data at the class level. Using OWL, we have defined class membership based solely on one or more identifiers, and therefore can yield logical equivalence between different data records. Class based representations mean that all instances will inherit the attributes of their type, and hence the challenge is to identify which identifiers in fact are equivalent. For instance, PubChem identifiers are unique, but several structures can map to CAS numbers. Using the logical framework of OWL it is possible to generate an inconsistency when two records are said to be different when they are in fact the same. More work is required in this regard to identify logically consistent identifier mappings.

Conversion of legacy data: There exists major challenge in creating ontologically structured knowledge from textual or semi-structured data. While DrugBank is a good resource for finding information about drug classes or drug targets, the meaning of this data is in free text rather than having been selected from controlled vocabularies. Our conversion is largely limited to highly regular fields such as FDA status or links to PubMed papers, and the remainder lies in the form of annotations (rather than class restrictions). Since DrugBank annotates their drugs using a shallow set of drug categories, we expect to further refine these into a nicely structured ontology and/or mapped to existing drug ontologies (i.e. ChEBI). There is a need for investigating new techniques to mine the large amount of textual information embedded as general descriptions, indications, toxicity, mechanism of action, absorption, dosage forms, among others into coherent structured knowledge.

Chemical conversion: Our plugin offers enormous flexibility in converting unstructured descriptors from sources other than PubChem. The configuration file can be modified so as to create a minimal knowledge base with only essential information, or can be used to map a wide variety of descriptors to ontological concepts, whether ours or their own. To maintain compatibility, users can specify a number of relationships (equivalence, subclass, type, sub-property) to concepts defined in our ontologies.

5 Conclusion

In this paper, we described a chemical knowledge representation for an OWL-based knowledge model. We integrate and query across PubChem, DrugBank and DBpedia in a way that is not possible using traditional database technologies. Indeed, by leveraging Semantic Web technologies, it becomes possible to integrate chemical information at differing levels of detail and granularity, opening new avenues for life science knowledge discovery.

References

1. Murray-Rust, P., Rzepa, H.S.: Chemical markup, XML and the World-Wide Web. 2. Information objects and the CMLDOM. J. Chem. Inf. Comput. Sci. 41, 1113–1123 (2001)
2. Feldman, H.J., Dumontier, M., Ling, S., Haider, N., Hogue, C.W.: CO: A chemical ontology for identification of functional groups and semantic comparison of small molecules. FEBS Lett. 579, 4685–4691 (2005)
3. Brooksbank, C., Cameron, G., Thornton, J.: The European Bioinformatics Institute's data resources: towards systems biology. Nucleic Acids Res. 33, 46–53 (2005)
4. Berners-Lee, T., Hendler, J., Lassila, O.: The Semantic Web. Scientific American (2001)
5. W3C: OWL Web Ontology Language Guide. In: Smith, M.K., Welty, C., McGuinness, D.L. (eds.): W3C Recommendation (2004)
6. Horrocks, I.: Applications of Description Logics: State of the Art and Research Challenges. In: Dau, F., Mugnier, M.-L., Stumme, G. (eds.) ICCS 2005. LNCS (LNAI), vol. 3596, pp. 78–90. Springer, Heidelberg (2005)
7. Grenon, P., Smith, B., Goldberg, L.: Biodynamic ontology: applying BFO in the biomedical domain. Stud. Health Technol. Inform. 102, 20–38 (2004)
8. Smith, B., Ceusters, W., Klagges, B., Kohler, J., Kumar, A., Lomax, J., Mungall, C., Neuhaus, F., Rector, A.L., Rosse, C.: Relations in biomedical ontologies. Genome Biol. 6, R46 (2005)
9. Horrocks, I., Patel-Schneider, P., Sattler, U., Parsia, B., Motik, B., Bechhofer, S., Calvanese, D., Giacomo, G.d., Lutz, C.: OWL 1.1 Specification (2006)
10. Wishart, D.S., Knox, C., Guo, A.C., Shrivastava, S., Hassanali, M., Stothard, P., Chang, Z., Woolsey, J.: DrugBank: a comprehensive resource for in silico drug discovery and exploration. Nucleic Acids Res. 34, 668–672 (2006)
11. Horridge, M., Drummond, N., Goodwin, J., Rector, A., Stevens, R., Wang, H.H.: The Manchester OWL Syntax. OWL Experiences and Design, Athens, Georgia (2006)
12. Motik, B., Grau, B.C., Sattler, U.: Structured Objects in OWL: Representation and Reasoning. In: 17th Int. World Wide Web Conference (WWW 2008), pp. 169–182. ACM Press, Beijing, China (2008)

Combining One-Class Classification Models Based on Diverse Biological Data for Prediction of Protein-Protein Interactions

José A. Reyes[1,2] and David Gilbert[1]

[1] Bioinformatics Research Centre, Department of Computing Science,
University of Glasgow, Glasgow, UK, G12 8QQ
{jareyes,drg}@dcs.gla.ac.uk
[2] Facultad de Ingeniería, Universidad de Talca, Chile

Abstract. This research addresses the problem of prediction of protein-protein interactions (PPI) when integrating diverse biological data. Gold Standard data sets frequently employed for this task contain a high proportion of instances related to ribosomal proteins. We demonstrate that this situation biases the classification results and additionally that the prediction of non-ribosomal based PPI is a much more difficult task. In order to improve the performance of this subtask we have integrated more biological data into the classification process, including data from mRNA expression experiments and protein secondary structure information. Furthermore we have investigated several strategies for combining diverse one-class classification (OCC) models generated from different subsets of biological data. The weighted average combination approach exhibits the best results, significantly improving the performance attained by any single classification model evaluated.

1 Introduction

The prediction of protein-protein interactions (PPI) has emerged recently as an important problem in the fields of Bioinformatics and Systems Biology due to the fact that many essential cellular processes such as signal transduction, transport, cellular motion and most regulatory mechanisms including gene regulatory process are mediated by this kind of interactions. High-throughput methods for the direct identification of protein-protein interactions have been developed including yeast two-hybrid screens (Y2H) [1,2] and mass spectrometry methods for protein complex identification [3,4]. Even though these high-throughput techniques can considerably increase the number of predicted PPI, in general the data obtained by these methods is often incomplete and suffers from high false-positive and false-negative rates [5]. In order to improve the accuracy and trustability of predicted protein interacting pairs various studies have been developed in the past years focused on the integration of diverse biological sources of information which could potentially incorporate new indirect clues related to protein interactions, demonstrating that the combined use of direct and indirect biological insights can improve the quality of predicted PPI.

A. Bairoch, S. Cohen-Boulakia, and C. Froidevaux (Eds.): DILS 2008, LNBI 5109, pp. 177–191, 2008.

The prediction of PPI has been commonly viewed as a classical binary classification problem where the aim is to predict whether any two proteins do or do not interact. Several traditional machine learning methods have been employed in the past for this specific task [6,7,8,9,10,11]. These methods employ supervised learning algorithms where the final objective is to generate a classification model from a gold standard reference set of positive (interacting pairs) and negative examples (non-interacting pairs). Two major drawbacks have been associated with this approach in the past: firstly imbalanced class problems where the number of positive examples (pairs of proteins which really interact) is much less than the number of negative ones. Secondly, while the selection of positive examples is based on trustable experimental techniques (i.e. small scale experiments), the set of negative examples is selected based on some assumptions, because there is no experimental method to find pairs of proteins which do not interact, which could introduce some bias in the classification results [12].

In the work presented in [13] we introduced the use of one-class classification (OCC) methods as a solution to these problems. OCC methods use feature information from only one of the classes (i.e. trustworthy positive examples in this case) in order to generate a classification model which consequently is independent of the kind of negative gold standard examples employed [14]. Additionally OCC methods are able to efficiently deal with highly imbalanced classification problems [15]. Among various OCC methods we evaluated the Parzen OCC density estimation approach clearly exhibited the best performance. Also in [13] we demonstrated that the Parzen OCC method performs competitively with those generated by conventional classifiers (i.e. Decision Trees, Support Vector Machines and Naive Bayes) and outperformed them in many situations. Additionally we reported that the performance of these conventional binary classification approaches is highly influenced by the quantity of negative examples used for training the respective models. This suggests that conventional classification models are more reliant on negative information (an untrustworthy set of negative PPI examples) than on positive information (experimentally corroborated PPI examples).

In this paper we address a potential new drawback which appears to affect the performance of the prediction of PPI. We focussed on the prediction of co-complex relationship in yeast, where the objective is to identify and characterize protein pairs which are members of the same protein complex. We found that the positive gold standard data set, which has been employed in the past in many related investigations, contains a high proportion of examples associated with interactions of ribosomal proteins. Here we demonstrate that this situation indeed biases the classification task, resulting for instance in an over-optimistic performance result. We divided the general classification task into two subtasks, the prediction of ribosomal and non-ribosomal PPI (i.e. dividing the gold standard set in two different sets). It can be seen that while the prediction of ribosomal PPI can be performed with high accuracy, the prediction of non-ribosomal PPI is a much difficult task.

Turning our attention exclusively into the subtask of prediction of non-ribosomal PPI we investigated some strategies in order to improve its performance. Firstly we considered the integration of more biological features to the classification process, related to mRNA expression and protein secondary structure information. Then we investigated and demonstrated that by combining the predictions of several Parzen OCC models induced from different subsets of biological data, it is possible to increment significantly the performance of prediction of non-ribosomal PPI.

The rest of the paper is organized as follows. In section 2 we demonstrate the effect associated with the high proportion of ribosomal PPI in the reference data set, and describe some methodological issues. Section 3 deals with the problem of prediction of non-ribosomal PPI, describing how new biological features are integrated to the classification process. In section 4 we investigate various strategies to combine independent Parzen OCC models generated employing different subsets of biological information. Finally section 5 concludes the paper.

2   Analysis of Positive Gold Standard Set

This research focusses on the prediction of co-complex protein pairs (pairs of proteins which are co-members of the same protein complex). In order to develop this classification task we need a reference data set or gold standard set containing positive (true interacting protein pairs) and negatives examples (non-interacting protein pairs). Although only positive examples are needed in order to train OCC methods, a set of negative ones is still required to obtain a comparable performance evaluation measure. Here we extend the data set we previously employed in [13] to consider a larger number of positive and negative examples. We follow the work in [6] to derive the positive gold standard set from the MIPS complex catalogue [16], and also the negative gold standard set which is related to protein pairs which are present in different cell localization and consequently are more likely not to interact. A similar reference data set has been employed before in [6,7,9,17]. The final data set we employed in this research includes $\sim$6,700 positive examples and $\sim$550,000 negative ones considering only examples where complete information for each one of the biological features were available.

Three different types of biological data were considered as features to develop our classification approach [1]:

- *mRNA expression,* the Pearson correlation is estimated for every protein pair considering two different studies the Rosetta compendium [18] and cell cycle time series analysis [19].
- *Functional similarity,* of protein pairs was estimated from the gene ontology (GO) [20] and the MIPS [16] functional catalog, obtaining two new numeric features. The assumption here is that proteins in the same complex tend to participate in the same biological processes.

---

[1] More details about how these features were estimated and encoded for the classification task can be found in [13].

- *Essentiality information,* was also used [16], assuming that is more expected that two proteins in the same complex are both essential or non essential but not a mixture of these two attributes.

Analyzing the composition of the positive gold standard set we found that a high proportion of these examples ($\sim$66%) are related to ribosomal protein pairs. This is because ribosomal protein complexes (cytoplasmic and mitochondrial) are the most numerous among all the different complexes included in the MIPS complex catalogue [16] which contain a large number of proteins. In this research we argue that this situation could considerably affect the performance of the classifiers, biasing the classifiers to mostly recognize interactions related to ribosomal proteins. In order to assess this situation we proceeded to divide our positive gold standard set in two subsets containing all ribosomal related PPI and all non-ribosomal related PPI respectively, generating at the same time two new classification subtasks related to the prediction of ribosomal and non-ribosomal PPI. The new positive gold standard sets contain $\sim$4,600 and $\sim$2,100 protein pairs respectively. We employed the same negative reference data set in both cases.

The performance of Parzen OCC approach was evaluated for the three situations considered above. More details about how the Parzen OCC approach works are given later in this section. Here we follow the same approach as in [13] to evaluate the performance of different classification methods. For the case of prediction of PPI we are specifically interested in evaluating the performance of the different classifiers under conditions of low false-positive rate, aiming to maximize the number of real interacting protein pairs predicted while minimizing the number of false-positive predicted ones. This is of especial interest for biologists working on the identification and validation of new PPI, because they can focus in the study of only the top ranked predicted PPI targets, instead of evaluating random protein pairs. Receiver Operator Characteristic (ROC) curves, which show the tradeoff between the false-positive rates and true-positive rates, were generated. But instead of calculating the area under the whole ROC curve or AUC score, here we consider the normalized area under the portion of the ROC curve related to the first 50 false-positive examples or AUC50 score. This measure has become a common and accepted performance measure for this specific task [10,11,13]. Additionally this performance measure is related to low values of false-positive rates and thus is more relevant in situations of severe class imbalance as in the case of prediction of PPI [21]. A ten fold cross validation procedure was performed for every evaluation in order to assess the variability of the models generated. The performance of the Parzen OCC classifier for the different tasks mentioned above is shown in Table 1. Several conventional classifiers were also included in this evaluation (Decision Trees, Support Vector Machines and Naive Bayes). For this a balanced class set was created using all the positive examples available and an equal size sample of negative examples randomly selected from the whole negative gold standard set. The WEKA machine learning library [22] was used to perform the experiments related to Decision Trees and Naive Bayes, while the evaluation of Support Vector Machines was carried out

Table 1. Performance of different classifiers measured as AUC50 scores. Three cases are evaluated: prediction considering all PPI in the positive gold standard set, prediction of ribosomal PPI and prediction of non-ribosomal PPI. AUC50 scores given as mean value and standard deviation (in brackets) based on a ten fold cross validation procedure.

Classifier	All PPI	ribosomal	non-ribosomal
Parzen OCC	0.5425 (0.0228)	0.7422 (0.0121)	0.1239 (0.0179)
Binary classifiers:			
Decision Trees	0.4916 (0.2902)	0.4808 (0.4486)	0.0439 (0.0280)
Naive Bayes	0.0064 (0.0021)	0.4710 (0.0202)	0.0207 (0.0105)
Support Vector Machines	0.2687 (0.0250)	0.5479 (0.1217)	0.0433 (0.0124)

using the MATLAB interface to the SVM-light toolbox [23]. The performance of conventional classifiers is also given in Table 1.

We observed a clear difference between the performance in the prediction of ribosomal and non-ribosomal PPI. In the case of prediction of ribosomal PPI the Parzen OCC approach exhibits a high performance of ~0,75 measured as an AUC50 score. The prediction of non-ribosomal PPI seems to be a more difficult task, here the performance of Parzen OCC approach is significantly reduced to only ~0.12 measured as an AUC50 score. Interestingly the performance in the situation when all PPI available in the positive gold standard are employed reach an AUC50 score of ~0,54 which is in-between the performance of both subtasks. The same behavior was observed when conventional classifiers were evaluated. the results show that the Parzen OCC approach clearly outperforms all conventional classification techniques for the different task evaluated, confirming our previous results reported in [13].

These results suggest that the performance obtained using the whole positive gold standard set is biased towards the prediction of ribosomal related PPI. The high performance exhibited in the prediction of the ribosomal PPI can be explained because they share common patterns in most of the biological features employed in the classification process, specifically those associated with functional similarity and mRNA expression based features. This is not the case when predicting non-ribosomal PPI which appears to be a much more difficult challenge and needs more attention by the scientific community in order to improve its performance. However a similar positive gold standard set derived from MIPS complex catalogue [16] has been employed in many studies related to the prediction of co-complex PPI [6,7,8,9,11,17,24]. The problem associated with the high proportion of ribosomal related proteins has not been previously reported or addressed according to the best of our knowledge. Furthermore in this paper we have focused on the task of prediction of non-ribosomal PPI and how to improve the performance of the Parzen OCC method for this task.

2.1 Parzen OCC Method

Here we briefly describe the main issues associated with the use of the Parzen OCC approach for the task of prediction of PPI. In this research we treated this as a OCC problem in the sense that only examples of one class, the positive interaction examples, are available and/or trustable, becoming the *target class*. These examples are used to generate a classification model which is used afterwards to discriminate between positive and negative examples also called the *outlier class*.

Every pair of proteins available in the gold standard set is represented by a vector X_i containing the information for the biological features considered here, and a label Y_i which can take two values depending on whether the proteins in the pair do really interact ($Y_i = 1$) or not ($Y_i = -1$). The dd_tools Matlab toolbox [25] was employed to develop the experiments associated with the application and evaluation of the Parzen OCC method. Here an independent Gaussian distribution is considered for each one of the T target objects (positive PPI examples) used for training a model. In order to classify a new object X the distance to all training objects is employed and a function $f(X)$ is estimated as:

$$f(X) = \sum_{i=1}^{T} \exp(-(X - X_i)^T h^{-2}(X - X_i)) \tag{1}$$

The smoothing parameter h, commonly called the *Parzen width*, is used here and is related to the width of a region R (in a Gaussian space) generated around each object in order to separate the target from outlier zones. The $f(X)$ value for new objects is then compared with a threshold θ and classified as a target if $f(X) \geq \theta$ or else as an outlier. In this research the threshold θ was set in a way that none of the positive examples employed to train the model is misclassified. The value of h can be varied in order to find an optimal performance related to the specific task conditions. The Parzen OCC classification model finally assigns a confidence value to each prediction.

3 Integration of Biological Information

3.1 mRNA Expression Integration

In order to improve the performance of prediction of non-ribosomal PPI we evaluated the effect of integrating more biological information into the classification process. The first approach developed was related to the integration of information associated with mRNA expression experiments. Here we explore the idea that m-RNA expression data obtained under different experimental conditions could give insights about different sets of new potential PPI. This is related to the identification of PPI sub-networks associated with cell adaptation to changing environments proposed and discussed in detail in [26]. We integrated the data generated in [27] related to yeast stress response. mRNA data previously employed in our study was related to yeast cell-cycle time series analysis [19] and the Rosetta compendium [18] which was related to gene mutations and chemical treatments. We evaluated the performance of the Parzen OCC method for this new data set following the same procedure as described in section 2. Initially we considered the case when all the biological features are integrated in a single data set which is then employed to generate and evaluate the performance of the Parzen OCC method. In order to evaluate the individual effect of the different mRNA expression data in the performance of the Parzen classifier. We also considered the cases where information related to only one of the mRNA expression experiments is employed. Finally we considered the situation where no mRNA expression data is employed. The results for all these are exhibited in Table 2 (middle column).

Table 2. Performance for diverse sets of biological data measured as AUC50 scores. AUC50 scores given as mean value and standard deviation (in brackets) based on a ten fold cross validation procedure

Description of data employed	mRNA Integration AUC50	Plus SS Integration AUC50
All mRNA expression data	0.1404 (0.0033)	0.2271 (0.0183)
Only Rosetta Compendium	0.1424 (0.0249)	0.2395 (0.0177)
Only Cell-Cycle	0.1859 (0.0208)	0.2344 (0.0146)
Only Stress response	0.2493 (0.0283)	0.2694 (0.0181)
No mRNA expression data	0.1249 (0.0220)	0.2656 (0.0238)

We observed that when all the data is employed together the performance of the Parzen OCC classifier is only slightly improved, reaching an AUC50 score of ~ 0.14 (compared with an AUC50 score of 0.1239 in the original situation as shown in Table 1). When data from only one mRNA expression experiment is employed we found a significant increment in the performance of the Parzen OCC method for the case of cell-cycle and stress response condition and a slight increment when using Rosetta experiments. The fact that models based on individual mRNA information perform better than the case when all data is integrated together suggests that the integration of features related to diverse mRNA expression conditions does not have a synergistic effect in the performance of the Parzen OCC method. On the contrary the integration of these features in a single data set seems to induce some kind of misclassification effect and consequently tends to reduce the overall performance. One possible explanation of this situation is that individual mRNA expression data sets (related to different experimental conditions) give different insights to the prediction problem. Moreover the classifier based on all the features together is not able to correctly discriminate between these situations. Finally the case when no mRNA information is employed exhibits a performance similar to the one obtained in the original situation described in section 2. Considering all these results we believe that it might be useful to investigate other ways to combine the information related to individual mRNA predictive models – see section 4.

3.2 Protein Secondary Structure Integration

Following the idea of integrating more biological information we investigated the use of protein secondary structure (SS) information. SS information has been employed in recent years for the characterization of protein-protein binding sites [28,29,30,31]. However these approaches consider only a reduced number of PPI which have been crystallized and are available in the Protein Data Bank (PDB) and additionally are focused exclusively on the interaction site region. In our approach we extend this idea to incorporate a larger number of PPI. To the best of our knowledge this is the first investigation associated with the use of secondary structure information for the prediction of PPI in a broad context.

In order to develop our approach instead of using 3D structure information we employed the whole linear protein sequence which is available for all yeast proteins. For each protein involved in our study we predicted the SS and relative solvent accessibility (RSA) for each residue employing the SSPRO program [32]. In this case SS is related to three possible classes for each residue, helix (H), strand (E) and the rest(C). RSA is associated with buried (b) or exposed (e) residues. Once SS ad RSA sequences have been predicted we faced the problem of how to generate features that could reflect some kind of relationship between SS and RSA for any two proteins. These features were then integrated into our general task of prediction of PPI and so were estimated for each instance included in the positive and negative PPI gold standard sets. A total of 13 features were generated as follows:

- *SS similarity:* Three features were generated based in the similarity of two SS sequences. Local and global alignments scores were estimated using the SSEA software [33]. Additionally we incorporated the common Edit Distance between them.
- *SS and RSA composition:* Four features were generated based on the SS and RSA composition following the work in [34]. For every protein a composition vector H,E,C,b,e containing the fraction of each residue type in the whole sequence was estimated. Then four similarity scores were calculated using dot product, cosine, Gaussian kernel and correlation between any two composition vectors.
- *Ratios:* Six features were generated based on the ratios of the composition of SS and RSA (measure this time as the number of residues of each type) and the total protein sequence length.

Firstly we evaluated the performance of the Parzen OCC method when only the 13 features based on SS and RSA information were employed. However the results (AUC50 scores) in this case were very poor (results not reported in this paper). Further we evaluated the effect of integrating these 13 features with the rest of the biological data previously employed. For this we used the same data set previously evaluated in section 3.1, incorporating the SS and RSA information for each of them. The results related to the performance of Parzen OCC approach when secondary information is integrated are shown in table 2 (left column).

We could see that the integration of secondary structure information has the effect of significantly incrementing the performance of the Parzen OCC approach in all situations (different subsets of biological data). This suggests that this type of information can indeed contribute to improving the performance of PPI prediction. Even though each of these features do not perform well when employed alone, it seems that integration with other types of biological data helps in the discrimination between positive and negative examples in the AUC50 region. Similar to the analysis developed in section 3.1 we again observed that models based on individual mRNA expression conditions perform better than when all biological information is employed together. This confirms our initial assumption that no synergistic effect is obtained when different mRNA expression data is utilized together. However in this case the effect seems to be less significant, which can be attributed to the presence of SS features.

Interestingly the strongest increment in the performance is shown in the case when no mRNA expression data is employed at all, more than doubling the performance of the original case. This suggests that the Parzen OCC model generated in this last configuration can give different insights to the problem of prediction of non-ribosomal PPI than those models based on individual mRNA expression information. This is also supported by the fact that SS based features contribute to improving the performance of every model based on individual mRNA data.

4 Combination of Diverse OCC Models

Based on the results obtained in the previous section, we further investigates the possibility of combining the predictions of different Parzen OCCC models in order to improve the performance of the prediction of non-ribosomal PPI. This exploits the idea of combining models that give us different insights to the problem of prediction of non-ribosomal PPI. Four models evaluated in section 3 were selected which could potentially satisfy this assumption. Three were based on individual mRNA expression experiments (without SS features) and one based on SS features with no mRNA information.

4.1 Diversity of Classification Models

By combining the predictions of different classifiers we aimed to improve the performance of the overall classification task [35]. This general approach is known under different names in the literature: *classifier ensembles, ensemble learning systems, mixture of experts, etc.* Other works [36,37,38] have shown that a good ensemble is only possible when the base classifiers perform diversely. This means correctly classifying and/or misclassifying different sets of objects. However diversity between classifiers can not ensure that there is an improvement in the overall performance. Without diversity there is no point in investigating the combination of diverse classification models.

In order to evaluate the diversity of the four selected classification methods we considered three general diversity measures commonly employed in the related literature: *Disagreement measure*, related to the degree of disagreement between two classifiers simply calculating the number of cases where one classifier is correct and the other is incorrect [39]; *Q statistics*, related in this case to the degree of similarity in the performance between two classifiers [40]; and *Kohavi-Wolpert variance*, which is associated with the variance derived from the decomposition formula of the classification error of a classifier [41]. To calculate these diversity measures for the four models selected in our approach we followed the general guidelines proposed in [36]. In our approach we are interested in the diversity of different classification models specifically in the AUC50 region (low false-positive rate values). Thus we adapted the diversity measures as follows. We considered exclusively the first "N" instances with the highest prediction confidence for each of the four Parzen OCC classifiers. Then we generated a unique list of instances integrating all selected sets. Finally instead of considering if an object is correctly or incorrectly classified by a classification model, we focussed on whether any object belonged or not to the highest confidence list of each model.

Estimates of these diversity measures are shown in Table 3. The results are given as mean value and standard deviation (in brackets) based on 10 fold cross validation (10FCV) procedure. These results were estimated using N=150, this value was selected arbitrarily considering that on each evaluation related to the 10FCV procedure around 200 positive examples are classified (non-ribosomal PPI gold standard set contains a total of ~2,100 instances). Diversity estimates employing N equals 100 and 200 were also calculated (results not included in the paper) exhibiting similar values. In the case of the Disagreement measure and Q statistics the average over all binary combinations of the four models selected was calculated. The table also shows the theoretical minimum and maximum values for each diversity measure considering the case when four models are combined. The Q statistic measure was normalized to have values between 0 and 1 (maximum diversity) following the approach in [37].

From the results in Table 3 it is possible to see that the four Parzen OCC classification models selected show a high diversity in all cases. This confirms our initial assumption that these models which were induced from diverse biological subsets of data give different insights into the problem of prediction of non-ribosomal PPI. Consequently this confirms our hypothesis that by combining their prediction it might be possible to improve the performance of the overall task.

Table 3. Variability of diverse models employed for combination process

Diversity measure	mean value	Min.	Max.
Disagreement	0.4946 (0.005)	0	1
Q statistic	0.5850 (0.022)	0	1
Kohavi-Wolpert variance	0.1855 (0.002)	0	0.25

4.2   Combination Strategies

In order to combine the predictions of the four Parzen OCC methods selected we investigated several strategies commonly employed in the literature. Each classifier in our ensemble assigns a predictive (or confidence) value to every object classified. These individual predictions were then combined in several ways in order to generate a single prediction score, which is employed for the final classification of diverse instances included in the test set. Four fixed combination rules were firstly investigated, which are related to the Mean, Median, Maximum and Product combination of the predictions of different classifiers. These approaches are fixed in the sense that it is not necessary to optimize any extra parameter(s). Additionally we investigated the weighted average combination approach, where different weights are assigned to each classifier prediction, and the finally prediction score was calculated by a linear combination of them [42].

In order to optimize the performance obtained by the weighted average combination approach (AUC50 score), we developed the following procedure. First constrain the sum of all weights to be equal to 1 (no negative weights were considered). Then evaluate the performance (AUC50 score) under different situations assigning different sets of weights to each classifier. For this we consider the whole range of possibilities, varying the weights assigned to each classifier between 0 and 1. Finally select the set of weights exhibiting the highest AUC50 score. The results derived using these combination strategies are shown in Table 4.

Table 4. Performance for diverse combination strategies measured as AUC50 scores. AUC50 scores given as mean value and standard deviation (in brackets) based on a ten fold cross validation procedure

Model combination strategy	AUC50
Mean combination	0.2897 (0.0218)
Median combination	0.2679 (0.0213)
Max combination	0.2226 (0.0234)
Product combination	0.3594 (0.0303)
Weighted average combination	0.3809 (0.0314)

We can see that most of the combination strategies produce an increment in the performance of the prediction of non-ribosomal PPI (with the exemption of the Maximum rule combination strategy), compared to the performances previously given in Table 2. The best performance was obtained when employing the weighted average combination approach. In this case an AUC50 score of over 0.38 was achieved, representing a significant increment in the performance of this task. The weights assigned to each classifier in the weighted average combination approach can be assigned a certain degree of importance. In the optimum situation achieved here the Parzen OCC model based on SS data without mRNA expression information was given the highest weight ($\sim$0.5), followed by the models based on mRNA expression associated with Stress response ($\sim$0.3), cell-cycle($\sim$0.15) and Rosetta compendium($\sim$0.05). The second best performance was achieved by the product combination approach with an AUC50 score of $\sim$0.36; interestingly this combination technique seems to perform well if the outcomes of individual classifiers are independent [43].

5 Conclusions

The research described in this paper addressed the problem of the prediction of co-complex PPI using the Parzen OCC method and integrating diverse kind of biological data. The positive gold standard set usually employed in this task contains a high proportion of ribosomal PPI. We have demonstrated that this situation introduces a bias in the classification task. We also showed that the subtask associated with the prediction of non-ribosomal PPI is a more difficult problem. This subtask has not received attention in the past, and our work is the first attempt to deal with this situation.

We focussed our efforts to improve the prediction of non-ribosomal PPI. We investigated the effect of in integrating new biological information into the process, based on data from mRNA expression experiments and protein secondary structure (SS) information. We have demonstrated that the integration of data from diverse mRNA expression experiments in a single data set has a negative effect in the performance of the Parzen OCC approach. There is no synergy effect in this case, and Parzen OCC models based on individual mRNA expression experiment outperform the one which integrates all the data. On the other hand the integration of protein secondary structure information results in a positive effect in the increment of performance of this predictive task. The performance of all of the models evaluated is improved when SS based features are incorporated into the classification process, including the case when no mRNA expression data is used. These results are very promising, and according to the best of our knowledge this is the first attempt to integrate this kind of information for the prediction of PPI.

Finally we investigated several strategies to combine predictions of different Parzen OCC models induced from diverse subsets of biological data. Four models were selected for this procedure, three based on individual mRNA expression experiments (without SS information) and one based on SS information (without mRNA expression data). These models exhibited a high degree of diversity in their predictions corroborating our assumption. We have demonstrated that it is possible to significantly improve the performance of the prediction of non-ribosomal PPI by combining the predictions of several Parzen OCC models. The weighted average combination approach exhibited the best performance, and also gave some insights regarding the relative importance of the different classifiers employed.

Acknowledgements

This work was supported by the Programme Alβan, the European Union Programme of High level Scholarships for Latin America, scholarship E04D034854CL.

References*

1. Uetz, P., Giot, L., Cagney, G., Mansfield, T.A., Judson, R.S., Knight, J.R., Lockshon, D., Narayan, V., Srinivasan, M., Pochart, P., Qureshi-Emili, A., Li, Y., Godwin, B., Conover, D., Kalbfleisch, T., Vijayadamodar, G., Yang, M., Johnston, M., Fields, S., Rothberg, J.M.: A comprehensive analysis of protein-protein interactions in saccharomyces cerevisiae. Nature 403, 623–627 (2000)
2. Ito, T., Chiba, T., Ozawa, R., Yoshida, M., Hattori, M., Sakaki, Y.: A comprehensive two-hybrid analysis to explore the yeast protein interactome. Proc. Natl. Acad. Sci. 98, 4569–4574 (2001)
3. Gavin, A.C., Bosche, M., Krause, R., Grandi, P., Marzioch, M., Bauer, A., Schultz, J., Rick, J.M., Michon, A.M., Cruciat, C.M., Remor, M., Hofert, C., Schelder, M., Brajenovic, M., Ruffner, H., Merino, A., Klein, K., Hudak, M., Dickson, D., Rudi, T., Gnau, V., Bauch, A., Bastuck, S., Huhse, B., Leutwein, C., Heurtier, M.A., Copley, R.R., Edelmann, A., Querfurth, E., Rybin, V., Drewes, G., Raida, M., Bouwmeester, T., Bork, P., Seraphin, B., Kuster, B., Neubauer, G., Superti-Furga, G.: Functional organization of the yeast proteome by systematic analysis of protein complexes. Nature 415, 141–147 (2002)
4. Ho, Y., Gruhler, A., Heilbut, A., Bader, G.D., Moore, L., Adams, S.L., Millar, A., Taylor, P., Bennett, K., Boutilier, K., Yang, L., Wolting, C., Donaldson, I., Schandorff, S., Shewnarane, J., Vo, M., Taggart, J., Goudreault, M., Muskat, B., Alfarano, C., Dewar, D., Lin, Z., Michalickova, K., Willems, A.R., Sassi, H., Nielsen, P.A., Rasmussen, K.J., Andersen, J.R., Johansen, L.E., Hansen, L.H., Jespersen, H., Podtelejnikov, A., Nielsen, E., Crawford, J., Poulsen, V., Srensen, B.D., Matthiesen, J., Hendrickson, R.C., Gleeson, F., Pawson, T., Moran, M.F., Durocher, D., Mann, M., Hogue, C.W.V., Figeys, D., Tyers, M.: Systematic identification of protein complexes in saccharomyces cerevisiae by mass spectrometry. Nature 415, 180–183 (2002)
5. von Mering, C., Krause, R., Snel, B., Cornell, M., Oliver, S.G., Fields, S., Bork, P.: Comparative assessment of large-scale data sets of protein-protein interactions. Nature 417, 399–403 (2002)
6. Jansen, R., Yu, H., Greenbaum, D., Kluger, Y., Krogan, N.J., Chung, S., Emili, A., Snyder, M., Greenblatt, J.F., Gerstein, M.: A bayesian networks approach for predicting protein-protein interactions from genomic data. Science 302, 449–453 (2003)
7. Lin, N., Wu, B., Jansen, R., Gerstein, M., Zhao, H.: Information assessment on predicting protein-protein interactions. BMC Bioinformatics 5(154) (2004)
8. Zhang, L., Wong, S., King, O., Roth, F.: Predicting co-complexed protein pairs using genomic and proteomic data integration. BMC Bioinformatics 5(38) (2004)
9. Lu, L.J., Xia, Y., Paccanaro, A., Yu, H., Gerstein, M.: Assessing the limits of genomic data integration for predicting protein networks. Genome Res. 15, 945–953 (2005)
10. Ben-Hur, A., Noble, W.S.: Kernel methods for predicting protein-protein interactions. Bioinformatics 21(suppl. 1), i38–i46 (2005)

11. Qi, Y., Bar-Joseph, Z., Klein-Seetharaman, J.: Evaluation of different biological data and computational classification methods for use in protein interaction prediction. Proteins: Structure, Function, and Bioinformatics 63, 490–500 (2006)
12. Ben-Hur, A., Noble, W.S.: Choosing negative examples for the prediction of protein-protein interactions. BMC Bioinformatics 7(S2) (2006)
13. Reyes, J.A., Gilbert, D.: Prediction of protein-protein interactions using one-class classification methods and integrating diverse data. Journal of Integrative Bioinformatics 4 (2007)
14. Tax, D.M.J., Duin, R.P.W.: Support vector data description. Machine Learning 54, 45–66 (2004)
15. Chawla, N.V., Japkowicz, N., Kotcz, A.: Editorial: special issue on learning from imbalanced data sets. SIGKDD Explorations 6, 1–6 (2004)
16. Mewes, H.W., Frishman, D., Guldener, U., Mannhaupt, G., Mayer, K., Mokrejs, M., Morgenstern, B., Munsterkotter, M., Rudd, S., Weil, B.: Mips: a database for genomes and protein sequences. Nucl. Acids Res. 30, 31–34 (2002)
17. Browne, F., Wang, H., Zheng, H., Azuaje, F.: An assessment of machine and statistical learning approaches to inferring networks of protein-protein interactions. Journal of Integrative Bioinformatics 3 (2006)
18. Hughes, T.R., Marton, M.J., Jones, A.R., Roberts, C.J., Stoughton, R., Armour, C.D., Bennett, H.A., Coffey, E., Dai, H., He, Y.D., Kidd, M.J., King, A.M., Meyer, M.R., Slade, D., Lum, P.Y., Stepaniants, S.B., Shoemaker, D.D., Gachotte, D., Chakraburtty, K., Simon, J., Bard, M., Friend, S.H.: Functional discovery via a compendium of expression profiles. Cell 102, 109–126 (2000)
19. Cho, R.J., Campbell, M.J., Winzeler, E.A., Steinmetz, L., Conway, A., Wodicka, L., Wolfsberg, T.G., Gabrielian, A.E., Landsman, D., Lockhart, D.J., Davis, R.W.: A genome-wide transcriptional analysis of the mitotic cell cycle. Mol. Cell 2, 65–73 (1998)
20. Ashburner, M., Ball, C.A., Blake, J.A., Botstein, D., Butler, H., Cherry, J.M., Davis, A.P., Dolinski, K., Dwight, S.S., Eppig, J.T., Harris, M.A., Hill, D.P., Issel-Tarver, L., Kasarskis, A., Lewis, S., Matese, J.C., Richardson, J.E., Ringwald, M., Rubin, G.M., Sherlock, G.: Gene ontology: tool for the unification of biology. The Gene Ontology Consortium. Nat. Genet. 25, 25–29 (2000)
21. Drummond, C., Holte, R.C.: Learning to live with false alarms. In: Workshop on Data Mining Methods for Anomaly Detection, Eleventh ACM SIGKDD International Conference on Knowledge Discovery and Data Mining (2005)
22. Witten, I.H., Frank, E.: Data Mining: Practical machine learning tools and techniques. Morgan Kaufmann, San Francisco (2005)
23. Joachims, T.: Making large-scale support vector machine learning practical. In: Schölkopf, B., Burges, C., Smola, A. (eds.) Advances in kernel methods: support vector learning, pp. 169–184. MIT Press, Cambridge (1999)
24. Van Berlo, R.J.P., Wessels, L.F., Ridder, D.D.E., Reinders, M.J.T.: Protein complex prediction using an integrative bioinformatics approach. J. Bioinform. Comput. Biol. 5, 839–864 (2007)
25. Tax, D.M.J.: Ddtools, the Data Description Toolbox for Matlab, http://www-ict.ewi.tudelft.nl/~davidt/dd_tools.html
26. Guo, Z., Li, Y., Gong, X., Yao, C., Ma, W., Wang, D., Li, Y., Zhu, J., Zhang, M., Yang, D., Wang, J.: Edge-based scoring and searching method for identifying condition-responsive protein protein interaction sub-network. Bioinformatics 23, 2121–2128 (2007)

27. Gasch, A.P., Spellman, P.T., Kao, C.M., Carmel-Harel, O., Eisen, M.B., Storz, G., Botstein, D., Brown, P.O.: Genomic expression programs in the response of yeast cells to environmental changes. Mol. Biol. Cell 11, 4241–4257 (2000)
28. Neuvirth, H., Raz, R., Schreiber, G.: Promate: a structure based prediction program to identify the location of protein-protein binding sites. J. Mol. Biol. 338, 181–199 (2004)
29. Hoskins, J., Lovell, S., Blundell, T.L.: An algorithm for predicting protein-protein interaction sites: Abnormally exposed amino acid residues and secondary structure elements. Protein Sci. 15, 1017–1029 (2006)
30. Guharoy, M., Chakrabarti, P.: Secondary structure based analysis and classification of biological interfaces: identification of binding motifs in protein protein interactions. Bioinformatics 23, 1909–1918 (2007)
31. Zhou, H.X., Qin, S.: Interaction-site prediction for protein complexes: a critical assessment. Bioinformatics 23, 2203–2209 (2007)
32. Cheng, J., Randall, A.Z., Sweredoski, M.J., Baldi, P.: SCRATCH: a protein structure and structural feature prediction server. Nucl. Acids Res. 33(suppl-2), W72–W76 (2005)
33. Fontana, P., Bindewald, E., Toppo, S., Velasco, R., Valle, G., Tosatto, S.C.E.: The SSEA server for protein secondary structure alignment. Bioinformatics 21, 393–395 (2005)
34. Cheng, J., Baldi, P.: A machine learning information retrieval approach to protein fold recognition. Bioinformatics 22, 1456–1463 (2006)
35. Dietterich, T.G.: Ensemble methods in machine learning. In: Kittler, J., Roli, F. (eds.) MCS 2000. LNCS, vol. 1857, pp. 1–15. Springer, Heidelberg (2000)
36. Kuncheva, L.I., Whitaker, C.J.: Measures of diversity in classifier ensembles and their relationship with the ensemble accuracy. Machine Learning 51, 181–207 (2003)
37. Tsymbal, A., Pechenizkiy, M., Cunningham, P.: Diversity in search strategies for ensemble feature selection. Information Fusion 6, 83–98 (2005)
38. Tang, E.K., Suganthan, P.N., Yao, X.: An analysis of diversity measures. Machine Learning 65, 247–271 (2006)
39. Ho, T.K.: The random subspace method for constructing decision forests. IEEE Transactions on Pattern Analysis and Machine Intelligence 20, 832–844 (1998)
40. Yule, G.U.: On the association of attributes in statistics. Philosophical Transactions of the Royal Society of London A(194), 257–319 (1900)
41. Kohavi, R., Wolpert, D.: Bias plus variance decomposition for zero-one loss functions. In: 13th International Conference on Machine Learning, pp. 275–283. Morgan Kaufmann, San Francisco (1996)
42. Kuncheva, L.I.: Combining Pattern Classifiers: Methods and Algorithms. Wiley-Interscience, Chichester (2004)
43. Duin, R.: The combining classifier: to train or not to train? In: 16th International Conference on Pattern Recognition, vol. 2, pp. 765–770 (2002)

Semi Supervised Spectral Clustering for Regulatory Module Discovery

Alok Mishra and Duncan Gillies

Imperial College, London SW7 2AZ, UK
alok.mishra@imperial.ac.uk

Abstract. We propose a novel semi-supervised clustering method for the task of gene regulatory module discovery. The technique uses data on dna binding as prior knowledge to guide the process of spectral clustering of microarray experiments. The microarray data from a set of repeat experiments are converted to an affinity, or similarity, matrix using a Gaussian function. We have investigated two methods to determine the optimal Gaussian variance for this purpose. The first method was based on a statistical measure of cluster coherence, and the second on optimising the number of constraints satisfied in the clustering process. The constraints, which were derived from dna-binding data, were used to adjust the affinity matrix to include known gene-gene interactions. Clusters were found using a spectrical clustering algorithm, and validated by using a biological significance score which was the proportion of gene pairs sharing a common transcription factor in the resulting clusters. Our results indicate that our technique can successfully leverage the information available in the dna-binding data. To the best of our knowledge this is a novel formulation for the purpose of gene module discovery.

1 Introduction

Complex functions of living cells in nature are carried out through the concerted activities of many genes and gene products which are organized into co-regulated sets also known as regulatory modules [1]. Understanding the organization of these sets of genes will provide insights into the cellular response mechanism under various conditions. Recently a considerable volume of data on gene activity, measured using several diverse techniques, has become widely available. By fusing this data using an integrative approach, we can try to unravel the regulation process at a more global level. Although an integrated model could never be as precise as one built from a small number of genes in controlled conditions, such global modelling can provide insights into higher processes where many genes are working together to achieve a task. Each of the different data types (microarray, dna-binding, protein-protein interaction and sequence data) provides a partial and noisy picture of the underlying actions. Hence there is a need to integrate them in order to obtain an improved and reliable picture of the whole process.

A considerable amount of work has been done by researchers [1,2,3] in order to integrate datasets to determine which groups of genes act together in modules.

A. Bairoch, S. Cohen-Boulakia, and C. Froidevaux (Eds.): DILS 2008, LNBI 5109, pp. 192–203, 2008.
© Springer-Verlag Berlin Heidelberg 2008

The techniques used to find these gene modules range from simple clustering to sophisticated statistical methods. We have used a novel formulation called *semi-supervised spectral clustering* which is a clustering algorithm in which supervision is provided in the form of constraints and clustering is done in spectral domain using the eigenvectors of an affinity matrix derived from the data. We have used Dna-binding (chromo-immunoprecipitation) datasets, which provide direct evidence of gene-transcription factor (TF) interaction, to derive the constraints in order to guide the clustering of microarray data. It is called a *semi-supervised* algorithm [4] because unlike traditional *supervised* clustering algorithms [5], it is not necessary to satisfy the constraints. Rather than acting as constraints to be *enforced*, they are used to *guide* the clustering process. Thus different datasets are not being merged *literally* but rather one is being used to guide the clustering of the other.

One of the advantages of our approach is that it can be used over the full range of bioinformatics data - both vectorial as well as non-vectorial. In recent years many similarity measures have been proposed for strings (dna sequences) [6] as well as graphs (protein-protein interaction) [7].

This paper starts with a description of background research in the field of supervised module discovery and a brief description of spectral clustering. This is followed by a detailed description of our algorithm, the datasets, their processing and the reasoning behind our choice of parameters. Lastly we have a section on discussion of results obtained followed by conclusions.

2   Background and Related Work

The concept of applying prior knowledge in the form of constraints to clustering algorithms is not new. Initial *supervised* algorithms were modifications of traditional ones and ensured that the resulting clusters had to satisfy certain constraints. One of the first papers in this area [5] proposed a constrained version of the famous *k-means* clustering algorithm by posing the problem in terms of minimum cost network flows. Their objective behind adding the constraints was to assign a certain minimum number of points to each cluster.

While such algorithms worked towards satisfying known constraints, other distance based clustering algorithms were developed in which the metric that a clustering algorithm uses in order to calculate distance between a pair of datapoints was modified by incorporating other sorts of information. These were the first *semi-supervised* clustering algorithms. They did not enforce the constraints but used them to provide guidance to the cluster formation process. This is the crucial difference between supervised and semi-supervised clustering algorithms. In the former approach the constraints are derived from known ground truth and have to be satisfied, whereas in the latter the constraints are additional sources of information but are considered noisy and hence not necessarily exactly correct. This is a characteristic of the dna-binding data that we use.

[8] proposed a distance metric that combines information from expression data and biological networks and uses it for clustering genes. They define a graph

distance function on a metabolic network derived from MIPS [9] and combine it with a correlation-based distance function for microarray gene expression measurements. They assigned equal weights to both the sources. The problem with this approach is that there is no justification for assigning equal weights to each of the sources. [10] developed a similar algorithm in which instead of combining the two information sources with equal weights, they used a shrinkage approach with the genes belonging to the same functional classes assigned zero distance (maximal similarity) and the rest of the genes using the distance calculated from the microarray data.

Our technique is likewise based on the concept of using the constraints obtained from one dataset in order to modify the similarity value that is obtained from another dataset. The key difference is that while all the previous work has used this principal to do clustering in some feature space, our technique uses the modified similarity values to cluster in spectral space (Spectral clustering). The field of spectral clustering itself was started by [11] who came up with the idea of constructing graph partitions using the eigenvectors of an adjacency matrix. It has generated a lot of interest in recent years [12,13] in clustering related research. One of the earlier applications of this technique to bioinformatics was by [14] who used it to cluster Gene Ontology terms to find sets of genes that might be functionally related. They used an information theoretic measure borrowed from text mining, where it had been used to calculate semantic similarities between words, to calculate the similarity values between the terms of the Gene Ontology.

2.1   Spectral Clustering

This is a clustering technique in which the eigenvectors of the *affinity* or *similarity* matrix with the highest eigenvalues are used to derive a clustering of given data points. Given a set of data points, some measure is used to calculate the pairwise similarity resulting in a similarity matrix. We can think of this similarity matrix as a graph with the data points as the nodes and each pairwise similarity value as the weight of an edge joining the pair. The clustering can now be defined as a graph partitioning problem where the edges between the points of a cluster have high weights while the edges between points belonging to different clusters are very low weights. The idea itself is not new but renewed interest recently has led to many new versions of this algorithm [12,13]. Our supervised version is a modification of [13] and is detailed in Section-3.

3   System and Methods

Our *primary* data set, on which the clustering is carried out, is the popular yeast microarray dataset [15] which was obtained by exposing yeast to various stress conditions. A similarity matrix was created using a Gaussian similarity function from this dataset. We selected only those genes that displayed a change of two fold in *at least one experiment*. There were 1062 genes fulfilling this criterion. The assumption behind this selection strategy is that the majority of genes which do not show much change in their expression levels during a process are unrelated to it. We used the log_2 of the ratio of the mean of Channel 2 (experimental expression) to the mean of Channel 1 (control expression) since this creates more symmetric distributions. The log-ratios are normalized so that each slide has zero mean and unit standard deviation.

The *secondary* dataset which we have used to guide the clustering process is the dna-binding dataset on yeast [16]. It was created using genome-wide location analysis techniques to determine the genomic occupancy of 203 DNA-binding transcriptional factors (TFs). In this dataset the likelihood of a particular TF binding to the promoter region of another gene is reported in terms of a confidence value (p-value). A lower p-value indicates higher confidence. In order to extract meaningful interactions we need to use some threshold on these reported p-values. We have used a range of p-value thresholds (from 0.1 to 0.0001) to indicate where there is significant TF binding. Since these are experimentally determined, each of these interactions is a constraint that we use for guiding the clustering process. In practice we investigated a range of p-value cut-offs on the dna-binding dataset, each corresponding to a certain set of constraints. This was to study the impact of *number* and *quality* of constraints on the biological significance of clustering.

As a next step, we used the p-value thresholds to convert the confidence value data into binary data. For example, if the p-value threshold is 0.001 then all values below this are considered as *definitely bound* and hence assigned a value of 1. The rest are assigned 0 (not bound). Therefore, our constraints are transformed into a $m \times n$ matrix where m is the number of genes and n is the number of TFs. This matrix is used to modify the similarity matrix that we obtain from the microarray data as indicated in Step-2 of Algorithm-1 and Figure-1. We only selected those genes that are common to both the datasets since some genes were missing in each.

3.1 Semi Supervised Spectral Clustering

We propose a semi-supervised form of the spectral clustering method, which is detailed in Algorithm-1 and Figure-1. We are clustering microarray data, hence the genes can be considered the nodes and the pairwise similarity values are calculated using a Gaussian affinity function. The reason behind using this affinity function is that it naturally encodes the local neighbourhood property and its value falls rapidly as the pairwise dissimilarity increases. Once we have this similarity matrix, we use the constraints derived from our secondary dataset to modify it. Since our constraints already encode our belief about potential interactions, we set each value in the similarity matrix to 1 (maximum similarity) if there is a 1 in corresponding constraints matrix. All other values are left unchanged as we have no information regarding them. The idea behind changing the values to represent maximum similarity is to give the algorithm the maximum incentive to keep them in the same cluster. The resulting matrix is the final similarity matrix that we use for spectral clustering (Steps 3-7). We calculate the normalized Laplacian and then find its eigenvalues. If we believe there are k clusters then eigenvectors corresponding to the k largest eigenvalues are chosen. These are then normalized and clustered using the k-means clustering algorithm. For all these integrated matrices, the k-means clustering of the eigenvectors was started from fixed centres. These 50 centres, each representing a cluster, were the genes encoding the TFs that had the highest numbers of dna-interactions in the dna-binding dataset.

Algorithm 1. Semi supervised Spectral clustering

Input: Microarray data matrix, Constraints matrix derived from DNA-binding data, width of the Gaussian(σ), number of clusters(k)

Output: k clusters comprising of all the genes in the microarray data matrix

1. Calculate the affinity matrix $K_{n \times n}$ from the microarray data matrix using Gaussian similarity function $k(\mathbf{x}, \mathbf{x}') = exp\left(-\frac{\|\mathbf{x}-\mathbf{x}'\|^2}{2\sigma^2}\right)$

2. Use the constraints to modify K, $K_{final} = K \oplus C$ where C is the constraints matrix. $K \oplus C$ implies that we set $K_{i,j} = 1$ where $C_{i,j} = 1$.

3. Calculate normalized Laplacian $L = D^{-1/2} K_{final} D^{-1/2}$ where D is the diagonal matrix with $d_{jj} = \sum_i d_{ji}$

4. Find the eigenvectors $v^1, v^2, \ldots, v^k$ corresponding to the largest k eigenvalues of L.

5. Use these eigenvectors as columns to get $V_{n \times k}$. Normalize it to have unit norm.

6. Cluster the points representing the rows of this matrix v_i using k-means algorithm into k clusters, $C_1, C_2, \ldots, C_k$.

7. Output clusters $A_1, A_2, \ldots, A_k$ such that $A_i = x_j \in C_i$

3.2   Parameter Selection

For any clustering algorithm, the most important decisions are the choice of the number of clusters and the free parameters. In our case, since the similarity among gene pairs is calculated using a Gaussian similarity function, the only free parameter is the width of the Gaussian, σ. For any unsupervised task of an exploratory nature, the *correct* number of clusters is data dependent. We chose to use 50 clusters in our experiments, based on earlier justifications by [1,17] which showed that the Saccharomyces Cerevisiae genome contains approximately 50 sets of functionally related genes. Both the authors have shown statistically that

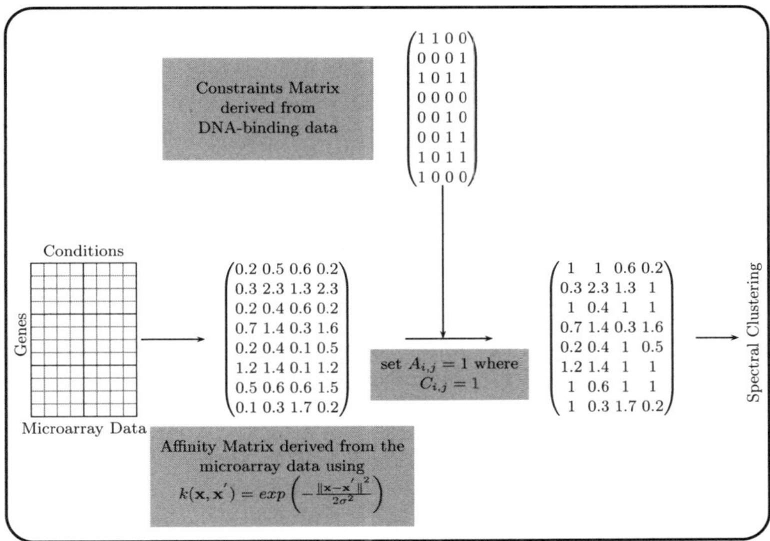

Fig. 1. Semi Supervised Spectral Clustering

this number provides a better fit to the underlying data distribution, compared to a higher or lower numbers of modules.

In order to determine the value of σ we initially used one of the most popular *internal* cluster quality validation index namely Dunn's Index [18]. Internal indices take a dataset and the resulting clustering and use information fully intrinsic to the data itself to assess the quality of clustering. This is different from *external* validation indices that use information independent of the dataset for validating the clustering. The underlying logic of using this to choose σ is to search for a value which results in the best quality clusters. We carried out this σ optimization without using the supervision step, clustering only the microarray dataset.

Dunn's index can be defined as

$$\text{Dunn index} = \min_{C_i \in C} \left(\min_{C_j \in C \setminus i} \left(\frac{dist(C_i, C_j)}{\max_{C_k \in C} diam(C_k)} \right) \right)$$

where $diam(C_k)$ is the maximum (complete) distance between two points within a cluster and $dist(C_i, C_j)$ is the minimum (single) distance between any two points in clusters C_i and C_j. We can observe that the value of this index is high if the inter-cluster separation is high compared to the largest cluster diameter. This corresponds to a fundamental objective of good clustering, namely to maximise the inter-cluster separation and minimise the intra-cluster distances. Hence *better* clustering will have *higher* values of this index. This index, though very easy to comprehend, can be quite unstable especially in presence of outliers.

We ran the spectral algorithm for various σ values. The range of σ values was determined as both the upper and lower extremes beyond which all the points resulted in a single cluster. For each σ value, we did 10 runs as Spectral Clustering depends on k-means which has random starting points. We also repeated the k-means algorithm twenty five times, each run being initialised randomly, and choose the best clustering with the minimum dispersion (within-cluster sum of squares). The results are shown in Figure-2(a) which shows the mean values along with standard deviation error bars. The x-axis uses a log-scale because of the spread of the data. Dunn's index has its maximum value (best clustering) at $\sigma = 0.003$. It is also worthwhile to note that the best quality clustering also has the least std. deviation.

As the standard deviation at many of the σ values was high, we investigated another independent method for estimating the best σ. For this, we took a very different approach relating to the use of the constraints. While adding supervision

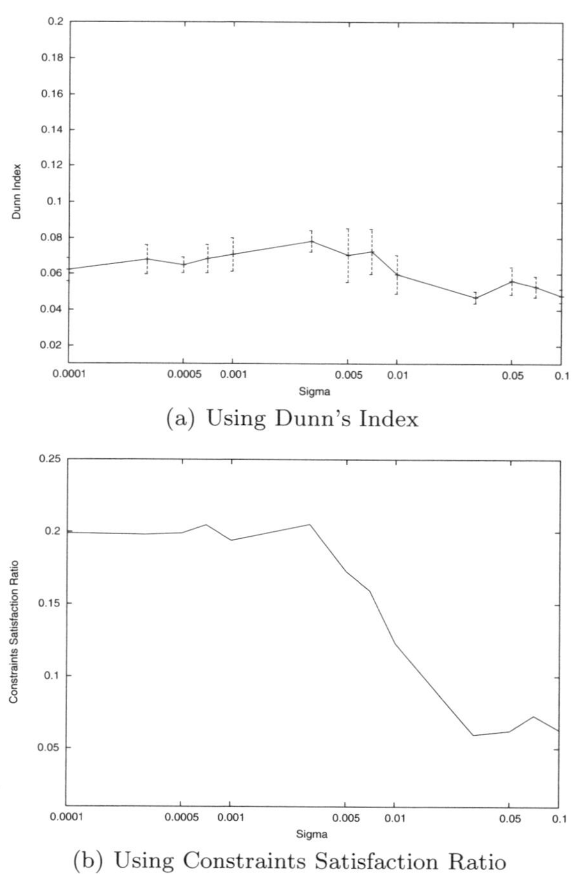

(a) Using Dunn's Index

(b) Using Constraints Satisfaction Ratio

Fig. 2. Sigma optimization using cluster quality tests

(constraints) we used a value of 1 (maximum similarity) irrespective of the value of σ (which determines the pairwise similarity values of genes). We believe that the optimum value of σ is the one for which the maximum number of constraints are satisfied. Therefore, we define Constraints Satisfaction Ratio (CSR) as

$$\text{CSR} = \left(\frac{\text{number of constraints satisfied}}{\text{number of constraints applied}} \right)$$

We used this index to confirm our choice of σ. As seen in Figure-2(b), the best value of sigma is again at at $\sigma = 0.003$. Please note that in this case we did not have to repeat runs of the clustering because as stated earlier, for all supervised matrices, the k-means clustering of the eigenvectors was started from fixed centres. These 50 centres (each representing a cluster) were the genes encoding the TFs that had the highest numbers of dna-interactions in the dna-binding dataset. Based on both (Dunn's index and CSR) results we can safely assume that the best clustering results are in this neighbourhood. We have not exhaustively searched the space of all possible σ values which can be done with a suitable optimization algorithm. We have used this σ values for all our further analysis.

4   Discussion

Evaluation of the results of our clustering algorithm requires careful consideration since there are no gold standards against which performance can be measured. The two prominent types of cluster validation measures are *internal* and *external* validation indices. As indicated earlier internal indices take a dataset and the resulting clustering and use information fully intrinsic to the data itself to assess the quality of clustering while external validation indices use information independent of the dataset for validating the clustering. We already saw the use of an internal validity measure for parameter (σ) selection. As they are fully dependent on the data itself, internal indices do not give any indication of the biological significance of resulting clusters.

There are various methods that have been used in the past for external validation most of which have used the information available in Gene Ontology. They calculate the statistical significance of various gene ontology terms in clusters. While this method gives us general ideas about which clusters might represent what functions, it doesn't allow us to functionally compare different clustering results *numerically*. Some attempts have been made to provide such a numerical index using mutual information and related concepts by [19,20].

We have evaluated the results of the semi-supervised clustering algorithm using our own external cluster validity index, which is based on the concept of counting gene pairs that have a *common* parent transcription Factor. We calculate a normalised count of such gene pairs in each cluster and use it to estimate the biological significance of the cluster. The gene pairs with a common transcription factor were not derived from the dna-binding dataset that we used for supervision but from an independently curated database, YEASTRACT [21] which has a collection of interactions between transcription factors and genes based on published research. In this database, the curators consider interaction to have occurred when there is change in the expression of the target gene owing to the deletion (or mutation) of the transcription factor-encoding gene. They also consider evidence based on TF binding to the promoter region of the target gene based on band-shift, footprinting or chromatin immunoprecipitation assays. They also describe potential associations but we have not considered them as we wanted our index to be as near to known facts as possible.

If N is total number of points in all the clusters and K is total number of clusters and if we define our clustering algorithm as an encoder $k = E(i)$ which assigns each data point to a cluster k then our Biological Significance Score, BSS is defined as

$$BSS = \frac{1}{K} \sum_{i=1}^{K} \frac{1}{\binom{N_i}{2}} \sum_{\substack{a \neq b \\ E(a)=E(b)=i}} C((PTF(a) \cap PTF(b)))$$

where

$$\binom{N_i}{2} = \frac{N_i * (N_i - 1)}{2} \, ,$$

$$N_i = \sum_{k=1}^{N} I(E(k) = i) \text{ and}$$

$$C(x) = \text{Cardinality of set x}$$

$$PTF(\mathbf{g}) = \text{set of TFs that are known to bind to gene } \mathbf{g}$$

Using this (BSS) score we were able to show that the algorithm can sucessfully use the information present in the dna-binding data. The original authors of the dna-binding dataset [16] have reported that they found the p-value of 0.001 to be the one which best represented known TF-dna interactions. It maximizes inclusion of legitimate TFs and minimizes false positives. Lower values were too strict and higher values found many false positives. We were able to show a similar trend with our score (BSS) when different p-value cut-offs were used for selecting the constraints from the dna-binding data. As discussed earlier in Section-3, p-values are used as cutoffs in order to get our constraints. A significant point to note is that these p-value cut-offs have a dual role. They determine the number of constraints as well as the quality of constraints. As the p-value cutoff is increased, the number of constraints also increases but a higher p-value also indicates lower confidence, hence the quality of the constraints falls. Table-1 shows the number of constraints corresponding to various p-value cut-offs. As a baseline we also calculated the value when no constraints are applied (p-value=10^{-6}).

From our results in Figure-3, we can see that with the addition of more constraints the cluster quality score improves till the p-value of 0.0005 and then gradually falls with increasing p-value after the peak. This signifies that when

Table 1. Number of Constraints with various p-value thresholds

p-value	Number of Constraints
0.0001	544
0.0005	846
0.001	1053
0.005	1959
0.01	2776
0.05	7407
0.1	12579

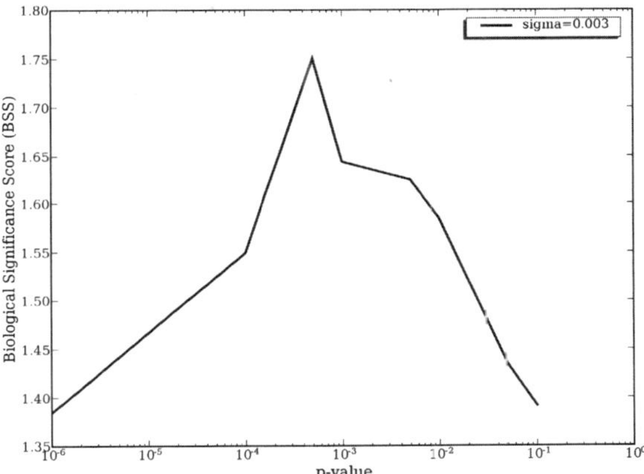

Fig. 3. Biological Significance with constraints

the number is larger than the optimum then the constraints represent noise and not-meaningful TF-gene interaction, and hence the clustering of microarray data is confused and the results get worse.

5   Conclusion

We have proposed a technique to integrate two diverse datasets where one is acting as a source of supervision on the clustering of the other. As part of this we have investigated two methods for detemining the best Gaussian kernel to obtain the affinity matrix from the data. Further, we have introduced a validation method which scores the resulting gene clusters by reference to a third type of data. By replicating the trend available in the DNA-binding data, our results demonstrate that the information available in it has been successfully incorporated in the combined matrix. However this does not necessarily prove that the resulting clusters are biologically more significant. Further work on validation using Gene-Ontology will help us to demonstrate the improvement in biological significance more convincingly.

We have used only dna-binding data as prior knowledge. Since our technique is quite generic, in future, we plan to extend it by using other sources as prior knowledge, for example the similarity derived from protein-protein interactions and the similarity between the promoter sequences of genes. In this paper we imposed an arbitrary cutoff on the binding data and thus converted indeterminate knowledge into definite knowledge. We would like to extend this study so that instead of creating definite constraints we consider the p-values as a similarity between the genes and then integrate the datasets. We can either assign weights in a shrinkage approach or develop better methods of weighing the importance of each dataset based on the data distributions in them.

One of the shortcomings of this research is that it is known that gene regulation is a very condition specific activity and hence the expression values that we observe are a result of regulation happening at one particular time. However there is no way to guarantee that the dna-binding data that we use represents the same time point in the regulation cycle as the microarray data. This is also a fundamental limitation of the the underlying experimental techniques, since microarrays themselves do not represent a single time point, but rather the integration of gene activity over a time period. Moreover knowledge about gene modules is not complete and this will hinder the validation process. Further research is required both in the measurement and analysis processes to improve our understanding of how genes interact.

References

1. Segal, E., Shapira, M., Regev, A., Pe'er, D., Botstein, D., Koller, D., Friedman, N.: Module networks: identifying regulatory modules and their condition-specific regulators from gene expression data. Nature Genetics 34(2), 166–176 (2003)
2. Bar-Joseph, Z., Gerber, G.K., Lee, T.I., Rinaldi, N.J., Yoo, J.Y., Robert, F., Gordon, D.B., Fraenkel, E., Jaakkola, T.S., Young, R.A., Gifford, D.K.: Computational discovery of gene modules and regulatory networks. Nature Biotechnology 21(11), 1337–1342 (2003)
3. Tanay, A., Sharan, R., Kupiec, M., Shamir, R.: Revealing modularity and organization in the yeast molecular network by integrated analysis of highly heterogeneous genomewide data. PNAS 101(9), 2981–2986 (2004)
4. Chapelle, O., Schölkopf, B., Zien, A. (eds.): Semi-Supervised Learning (Adaptive Computation and Machine Learning). MIT Press, Cambridge (2006)
5. Bradley, P.S., Bennett, K.P., Demiriz, A.: Constrained k-means clustering
6. Vert, J.-P., Thurman, R., Noble, W.S.: Kernels for gene regulatory regions. In: Weiss, Y., Schölkopf, B., Platt, J. (eds.) Advances in Neural Information Processing Systems 18, vol. 18, pp. 1401–1408. MIT Press, Cambridge (2006)
7. Kondor, R.I., Lafferty, J.D.: Diffusion kernels on graphs and other discrete input spaces. In: ICML, pp. 315–322 (2002)
8. Hanisch, D., Zien, A., Zimmer, R., Lengauer, T.: Co-clustering of biological networks and gene expression data. Bioinformatics 18 (suppl. 1) (2002)
9. Mewes, H.W., Amid, C., Arnold, R., Frishman, D., Gueldener, U., Mannhaupt, G., Muensterkoetter, M., Pagel, P., Strack, N., Stuempflen, V., Warfsmann, J., Ruepp, A.: Mips: analysis and annotation of proteins from whole genomes. Nucleic Acids Res. 32 Database issue (January 2004)

10. Huang, D., Pan, W.: Incorporating biological knowledge into distance-based clustering analysis of microarray gene expression data. Bioinformatics 22(10), 1259–1268 (2006)
11. Donath, W.E., Hoffman, A.J.: Lower bounds for the partitioning of graphs. IBM J. Res. Dev 17(5), 420–425 (1973)
12. Shi, J., Malik, J.: Normalized cuts and image segmentation. IEEE Transactions on Pattern Analysis and Machine Intelligence 22(8), 888–905 (2000)
13. Ng, A.Y., Jordan, M.I., Weiss, Y.: On spectral clustering: Analysis and an algorithm. In: NIPS, pp. 849–856 (2001)
14. Speer, N., Frlich, H., Spieth, C., Zell, A.: Functional grouping of genes using spectral clustering and gene ontology. In: Proceedings of the IEEE International Joint Conference on Neural Networks, pp. 298–303. IEEE Computer Society Press, Los Alamitos (2005)
15. Gasch, A.P., Spellman, P.T., Kao, C.M., Carmel-Harel, O., Eisen, M.B., Storz, G., Botstein, D., Brown, P.O.: Genomic expression programs in the response of yeast cells to environmental changes. Mol. Biol. Cell 11(12), 4241–4257 (2000)
16. Harbison, C.T., Gordon, B.D., Lee, T.I., Rinaldi, N.J., Macisaac, K.D., Danford, T.W., Hannett, N.M., Tagne, J.B., Reynolds, D.B., Yoo, J., Jennings, E.G., Zeitlinger, J., Pokholok, D.K., Kellis, M., Rolfe, A.P., Takusagawa, K.T., Lander, E.S., Gifford, D.K., Fraenkel, E., Young, R.A.: Transcriptional regulatory code of a eukaryotic genome. Nature 431(7004), 99–104 (2004)
17. Ihmels, J., Friedlander, G., Bergmann, S., Sarig, O., Ziv, Y., Barkai, N.: Revealing modular organization in the yeast transcriptional network. Nature Genet. 31, 370–377 (2002)
18. Dunn, J.: Well-separated clusters and optimal fuzzy partitions. Journal of Cybernetics 4, 95–104 (1974)
19. Gibbons, F.D., Roth, F.P.: Judging the Quality of Gene Expression-Based Clustering Methods Using Gene Annotation. Genome Res. 12(10), 1574–1581 (2002)
20. Gat-Viks, I., Sharan, R., Shamir, R.: Scoring clustering solutions by their biological relevance. Bioinformatics 19(18), 2381–2389 (2003)
21. Teixeira, M.C., Monteiro, P., Jain, P., Tenreiro, S., Fernandes, A.R., Mira, N.P., Alenquer, M., Freitas, A.T., Oliveira, A.L., Sa-Correia, I.: The YEASTRACT database: a tool for the analysis of transcription regulatory associations in Saccharomyces cerevisiae. Nucl. Acids Res. 34(1), 446–451 (2006)

UniProtKB/Swiss-Prot: New and Future Developments

Amos Bairoch

Swiss-Prot group at the Swiss Institute of Bioinformatics and
Department of Structural Biology and Bioinformatics at the University of Geneva
`Amos.bairoch@isb-sib.ch`

The Swiss-Prot knowledgebase [1] was created in 1986. It is now the cornerstone of the UniProt consortium [2] efforts that aims to provide to life scientists a wide range of information concerning proteins.

UniProtKB/Swiss-Prot provides concise, but thorough, descriptions of a non-redundant set of proteins, including their function, domain structure, post-translational modifications and variants. Swiss-Prot is tightly integrated with other databases, allowing the user to move seamlessly from sequence-based information to related information such as a protein's three-dimensional structure or its coding sequence. High quality manual annotation is what makes Swiss-Prot so useful to its academic and industrial users. Its unparalleled level of annotation is the fruit of manual curation by highly qualified biologists, who use their understanding of biology and the vast amount of information available in the scientific literature to provide accurate descriptions of each protein's features.

The driving factors in the development of the knowledgebase are:

- To constantly add new entries;
- To regularly update existing entries by adding newly available data;
- To provide new type of information;
- To standardize existing information to facilitate their retrieval and programmatic access by bioinformatics tools.

In 2007 we added 80829 entries to UniProtKB/Swiss-Prot thus reaching a total of 333445 entries. This was by far, the biggest increase in number of entries since the inception of the knowledgebase. At the same time we updated a huge number of existing entries, thus adding new annotations and cross-references to the existing corpus of knowledge.

In terms of new type of information and of standardization efforts, many things were achieved in 2007. The full list and precise description of these changes are listed in a web page which is updated at each tri-weekly release (`http://www.expasy.org/sprot/relnotes/sp_news.html`). We will only highlight here three significant developments.

a) We have introduced a new line type (PE which stands for Protein Existence) to indicate the evidences for the existence of a given protein. 5 levels of evidence have been defined:

A. Bairoch, S. Cohen-Boulakia, and C. Froidevaux (Eds.): DILS 2008, LNBI 5109, pp. 204–206, 2008.
© Springer-Verlag Berlin Heidelberg 2008

1. evidence at protein level (e.g. partial Edman sequencing, clear identification by mass spectrometry);
2. evidence at transcript level (e.g. Northern blot);
3. inferred by homology (strong sequence similarity to known proteins in related species);
4. predicted;
5. uncertain (e.g. dubious sequences that could be the erroneous translation of a pseudogene).

The full list of criteria that are used to assign the different levels is described in a new document file **pe_criteria.txt**.

b) We have structured the comment line topic SUBCELLULAR LOCATION in order to improve the consistency of annotation and to allow parsing of its content. A new document file **subcell.txt** lists the controlled vocabularies used in this topic, their definitions and further information such as synonyms or relevant GO terms.

c) We have added cross-references to 15 external resources, added links to Wikipedia and changed the format of cross-references to PDB to indicate the resolution of structures that were determined by X-ray crystallography or electron microscopy.

We plan to carry out quite a number of changes in the coming months. The most important one in term of its impact on all our users is a complete redefinition of the description (DE) lines. The UniProtKB description lines list protein names in a computer parsable format, but currently with a minimal amount of structure. Consistent nomenclature is indispensable for communication, literature searching and entry retrieval. The protein names provided in the description lines of UniProtKB/Swiss-Prot are widely used by life scientists and often propagated during the annotation of new genomic sequences. For these reasons we intend to structure the UniProtKB DE lines more explicitly: we will introduce two categories (recommended and alternative), as well as several subcategories, of protein names.

Example in the current format:

```
DE   Interleukin-2 precursor (IL-2) (T-cell growth factor) (TCGF)
DE   (Aldesleukin).
```

Example in the new format:

```
DE   RecName: Full=Interleukin-2;
DE            Short=IL-2;
DE   AltName: Full=T-cell growth factor;
DE            Short=TCGF;
DE   AltName: INN=Aldesleukin;
DE   Flags: Precursor;
```

This development along with the many others that are planned contribute to our mission which is to provide the scientific community with a comprehensive, high-quality and freely accessible resource of protein sequence and functional information.

References

[1] Bairoch, A., Boeckmann, B., Ferro, S., Gasteiger, E.: juggling between evolution and stability. Briefings Bioinform. 5, 39–55 (2004)
[2] UniProt Consortium: The universal protein resource (UniProt). Nucleic Acids Res. 36, D190–D195 (2008)

EBI Proteomics Services

Henning Hermjakob

European Bioinformatics Institute, Wellcome Trust Genome Campus, Hinxton, UK
hhe@ebi.ac.uk

Abstract. We will provide an overview of recent developments in the tools and resources provided by the EBI Proteomics Services Team.

The IntAct molecular interaction database (http://www.ebi.ac.uk/intact)[1] now offers 163.000 curated binary molecular interactions. Using the Distributed Annotation System (DAS)[2] and DASTY (http://www.ebi.ac.uk/dasty), we are currently enhancing the IntAct molecular view through direct integration with external resources, in particular UniProt and ChEBI. We will provide a detailed view of the process and our experience using DAS for data integration and robust data maintenance.

The PRIDE proteomics identifications database (http://www.ebi.ac.uk/pride) [3] provides more than 500.000 protein identifications, supported by more than 3 million identified peptides. We will report on recent improvements to data submission system and interface, in particular the visualization of PRIDE peptides through DAS, as well as PRIDE protein sets on Reactome pathways.

A major challenge not only, but particularly in proteomics data resources is the efficient management of consistent protein identifiers and controlled vocabularies. We will describe two major tools to address these tasks, the PICR Protein Identifier Cross-Referencing service (http://www.ebi.ac.uk/Tools/picr), and the OLS Ontology Lookup Service (http://www.ebi.ac.uk/ols). PICR provides high quality mapping between protein identifier namespaces, based on a database of more than 15 million unique protein sequences and associated identifiers. OLS provides efficient access to currently almost 700.000 terms from 60 controlled vocabularies in OBO format. Both systems provide interactive as well as web service access, and are publicly available without restrictions.

References

1. Kerrien, S., Alam-Faruque, Y., Aranda, B., et al.: IntAct–open source resource for molecular interaction data. Nucleic Acids Res. 35, 561–565 (2007)
2. Prlic, A., Down, T.A., Kulesha, E., et al.: Integrating sequence and structural biology with DAS. BMC Bioinformatics 8, 333 (2007)
3. Jones, P., Cote, R.G., Cho, S.Y., et al.: PRIDE: new developments and new datasets. Nucleic Acids Res. 36, 878–883 (2008)

A. Bairoch, S. Cohen-Boulakia, and C. Froidevaux (Eds.): DILS 2008, LNBI 5109, p. 207, 2008.

Bio-ontologies Tutorial

Olivier Dameron and Julie Chabalier

EA-3888, IFR 140, Université Européenne de Bretagne, Faculté de médecine
35043 Rennes, France

1 Description

We organize a three hours tutorial about bio-ontologies and OWL-DL during
the DILS'2008 conference.

The first part will present the motivations for using ontologies in the context of bioinformatics. Likewise, it will present the reference ontologies of the
domain (Gene Ontology, BioPAX, ...) and their typical use (e.g. Gene Ontology
Annotation).

The second part will be more hands-on oriented and will focus on the basic
principles for creating a simple ontology. We will cover the creation of subclasses
and of relations. At the time of the break, we will have demonstrated the expressivity of RDFS.

The third part will continue with the hands-on approach to cover OWL features such as conjunction and disjunction, negation, and existential and universal
constraints. We will demonstrate the associated reasoning capabilities in a bioinformatics' context. We will show how these capabilities can be used for maintaining a curated version of an ontology, as well as for enriching data processing
with symbolic reasoning. In order to cover some of the OWL-DL "peculiarities",
this part is organized as a series of thought-provoking situations where the initial result of classification is not what one would assume, before we demonstrate
what the misleading assumption was and how to overcome it.

2 Target Audience

This tutorial is oriented to attendees from both biological or computer-science
background, with no prior knowledge of ontologies. The third part can be of
interest for attendees focusing on reasoning.

The whole tutorial will be conducted in a semi-interactive way. All the technical steps will be demonstrated live (and rather slowly) so that attendees can
choose either to follow along with their personal laptop, or just to watch.

A. Bairoch, S. Cohen-Boulakia, and C. Froidevaux (Eds.): DILS 2008, LNBI 5109, p. 208, 2008.
© Springer-Verlag Berlin Heidelberg 2008

Printing: Mercedes-Druck, Berlin
Binding: Stein+Lehmann, Berlin

Lecture Notes in Bioinformatics

Vol. 5109: A. Bairoch, S. Cohen-Boulakia, C. Froidevaux (Eds.), Data Integration in the Life Sciences. XIII, 209 pages. 2008.

Vol. 5054: J. Fisher (Ed.), Formal Methods in Systems Biology. VII, 139 pages. 2008.

Vol. 4983: I. Măndoiu, R. Sunderraman, A. Zelikovsky (Eds.), Bioinformatics Research and Applications. XIX, 510 pages. 2008.

Vol. 4955: M. Vingron, L. Wong (Eds.), Research in Computational Molecular Biology. XVI, 480 pages. 2008.

Vol. 4780: C. Priami (Ed.), Transactions on Computational Systems Biology VIII. VII, 103 pages. 2007.

Vol. 4774: J.C. Rajapakse, B. Schmidt, L.G. Volkert (Eds.), Pattern Recognition in Bioinformatics. XVIII, 410 pages. 2007.

Vol. 4751: G. Tesler, D. Durand (Eds.), Comparative Genomics. IX, 193 pages. 2007.

Vol. 4695: M. Calder, S. Gilmore (Eds.), Computational Methods in Systems Biology. X, 249 pages. 2007.

Vol. 4689: K. Li, X. Li, G.W. Irwin, G. He (Eds.), Life System Modeling and Simulation. XIX, 561 pages. 2007.

Vol. 4645: R. Giancarlo, S. Hannenhalli (Eds.), Algorithms in Bioinformatics. XIII, 432 pages. 2007.

Vol. 4643: M.-F. Sagot, M.E.M.T. Walter (Eds.), Advances in Bioinformatics and Computational Biology. XII, 177 pages. 2007.

Vol. 4544: S. Cohen-Boulakia, V. Tannen (Eds.), Data Integration in the Life Sciences. XI, 282 pages. 2007.

Vol. 4532: T. Ideker, V. Bafna (Eds.), Systems Biology and Computational Proteomics. IX, 131 pages. 2007.

Vol. 4463: I.I. Măndoiu, A. Zelikovsky (Eds.), Bioinformatics Research and Applications. XV, 653 pages. 2007.

Vol. 4453: T. Speed, H. Huang (Eds.), Research in Computational Molecular Biology. XVI, 550 pages. 2007.

Vol. 4414: S. Hochreiter, R. Wagner (Eds.), Bioinformatics Research and Development. XVI, 482 pages. 2007.

Vol. 4366: K. Tuyls, R.L. Westra, Y. Saeys, A. Nowé (Eds.), Knowledge Discovery and Emergent Complexity in Bioinformatics. IX, 183 pages. 2007.

Vol. 4360: W. Dubitzky, A. Schuster, P.M.A. Sloot, M. Schröder, M. Romberg (Eds.), Distributed, High-Performance and Grid Computing in Computational Biology. X, 192 pages. 2007.

Vol. 4345: N. Maglaveras, I. Chouvarda, V. Koutkias, R. Brause (Eds.), Biological and Medical Data Analysis. XIII, 496 pages. 2006.

Vol. 4316: M.M. Dalkilic, S. Kim, J. Yang (Eds.), Data Mining and Bioinformatics. VIII, 197 pages. 2006.

Vol. 4230: C. Priami, A. Ingólfsdóttir, B. Mishra, H. Riis Nielson (Eds.), Transactions on Computational Systems Biology VII. VII, 185 pages. 2006.

Vol. 4220: C. Priami, G. Plotkin (Eds.), Transactions on Computational Systems Biology VI. VII, 247 pages. 2006.

Vol. 4216: M. R. Berthold, R.C. Glen, I. Fischer (Eds.), Computational Life Sciences II. XIII, 269 pages. 2006.

Vol. 4210: C. Priami (Ed.), Computational Methods in Systems Biology. X, 323 pages. 2006.

Vol. 4205: G. Bourque, N. El-Mabrouk (Eds.), Comparative Genomics X, 231 pages. 2006.

Vol. 4175: P. Bücher, B.M.E. Moret (Eds.), Algorithms in Bioinformatics. XII, 402 pages. 2006.

Vol. 4146: J.C. Rajapakse, L. Wong, R. Acharya (Eds.), Pattern Recognition in Bioinformatics. XIV, 186 pages. 2006.

Vol. 4115: D.-S. Huang, K. Li, G.W. Irwin (Eds.), Computational Intelligence and Bioinformatics, Part III. XXI, 803 pages. 2006.

Vol. 4075: U. Leser, F. Naumann, B. Eckman (Eds.), Data Integration in the Life Sciences. XI, 298 pages. 2006.

Vol. 4070: C. Priami, X. Hu, Y. Pan, T.Y. Lin (Eds.), Transactions on Computational Systems Biology V. IX, 129 pages. 2006.

Vol. 4023: E. Eskin, T. Ideker, B. Raphael, C. Workman (Eds.), Systems Biology and Regulatory Genomics. X, 259 pages. 2007.

Vol. 3939: C. Priami, L. Cardelli, S. Emmott (Eds.), Transactions on Computational Systems Biology IV. VII, 141 pages. 2006.

Vol. 3916: J. Li, Q. Yang, A.-H. Tan (Eds.), Data Mining for Biomedical Applications. VIII, 155 pages. 2006.

Vol. 3909: A. Apostolico, C. Guerra, S. Istrail, P.A. Pevzner, M. Waterman (Eds.), Research in Computational Molecular Biology. XVII, 612 pages. 2006.

Vol. 3886: E.G. Bremer, J. Hakenberg, E.-H.(S.) Han, D. Berrar, W. Dubitzky (Eds.), Knowledge Discovery in Life Science Literature. XIV, 147 pages. 2006.

Vol. 3745: J.L. Oliveira, V. Maojo, F. Martín-Sánchez, A.S. Pereira (Eds.), Biological and Medical Data Analysis. XII, 422 pages. 2005.

Vol. 3737: C. Priami, E. Merelli, P. Gonzalez, A. Omicini (Eds.), Transactions on Computational Systems Biology III. VII, 169 pages. 2005.

Vol. 3695: M. R. Berthold, R.C. Glen, K. Diederichs, O. Kohlbacher, I. Fischer (Eds.), Computational Life Sciences. XI, 277 pages. 2005.

Vol. 3692: R. Casadio, G. Myers (Eds.), Algorithms in Bioinformatics. X, 436 pages. 2005.

Vol. 3680: C. Priami, A. Zelikovsky (Eds.), Transactions on Computational Systems Biology II. IX, 153 pages. 2005.

Vol. 3678: A. McLysaght, D.H. Huson (Eds.), Comparative Genomics. VIII, 167 pages. 2005.

Vol. 3615: B. Ludäscher, L. Raschid (Eds.), Data Integration in the Life Sciences. XII, 344 pages. 2005.

Vol. 3594: J.C. Setubal, S. Verjovski-Almeida (Eds.), Advances in Bioinformatics and Computational Biology. XIV, 258 pages. 2005.

Vol. 3500: S. Miyano, J. Mesirov, S. Kasif, S. Istrail, P.A. Pevzner, M. Waterman (Eds.), Research in Computational Molecular Biology. XVII, 632 pages. 2005.

Vol. 3388: J. Lagergren (Ed.), Comparative Genomics. VII, 133 pages. 2005.

Vol. 3380: C. Priami (Ed.), Transactions on Computational Systems Biology I. IX, 111 pages. 2005.

Vol. 3370: A. Konagaya, K. Satou (Eds.), Grid Computing in Life Science. X, 188 pages. 2005.

Vol. 3318: E. Eskin, C. Workman (Eds.), Regulatory Genomics. VII, 115 pages. 2005.

Vol. 3240: I. Jonassen, J. Kim (Eds.), Algorithms in Bioinformatics. IX, 476 pages. 2004.

Vol. 3082: V. Danos, V. Schachter (Eds.), Computational Methods in Systems Biology. IX, 280 pages. 2005.

Vol. 2994: E. Rahm (Ed.), Data Integration in the Life Sciences. X, 221 pages. 2004.

Vol. 2983: S. Istrail, M.S. Waterman, A. Clark (Eds.), Computational Methods for SNPs and Haplotype Inference. IX, 153 pages. 2004.

Vol. 2812: G. Benson, R.D.M. Page (Eds.), Algorithms in Bioinformatics. X, 528 pages. 2003.

Vol. 2666: C. Guerra, S. Istrail (Eds.), Mathematical Methods for Protein Structure Analysis and Design. XI, 157 pages. 2003.